2ND EDITION

BOX OF ROCKS

A Left/Right Brain Guide to...

ROCK COLLECTING & CRYSTAL HEALING

S. Jay Wexler

"The best stone is always the next one found" –SJW

ACKNOWLEDGEMENTS

As always, I give my thanks to Lesley Wexler for continuing to be "Mrs. Rock Guy", for supporting my work, designing this book, housing the ever-expanding collection, and generally being a wonderful person. She is my favorite rock. Also, to our boys who inspire me every day with their own endeavors, wisdoms, witticism, and who put up with my sense of humor. Your groans make it all worthwhile.

Published by Wires, Pliers & Stones
©2022 by S. Jay Wexler
All rights reserved.

Written by S. Jay Wexler
Design and Layout by aha!, Inc.
Libertyville, IL 60048
Photographs by S. Jay Wexler

ISBN 979-8-218-03790-1

TABLE OF CONTENTS

WHAT IS THIS GUIDE AND HOW DO I USE IT?

Welcome to our family collection of rocks, minerals, gemstones and crystals! We created this new edition of this guide to organize as much pertinent information on the topic as we could. We are not professional geologists nor gemologists. We are simply a family of enthusiastic and inquisitive collectors like so many other people around the world. Since we are often seeking references to answer questions about the properties or composition of a specific stone, we decided to establish a repository for our cumulative information. This book serves as both a personal catalogue of our own collection as well as something that we hope others will enjoy or find useful.

Each entry page has mineralogical information that may be further illuminated in the indexes at the end of the book. There is also an index of healing properties and references for metaphysical uses for rocks and crystals. While this aspect of the guide is in no way scientific in basis, we believe the information should be available for those who wish to consider the many possible benefits. (**Any health information in this book should not be considered medical advice in any way and should not replace treatment or diagnosis by a medical professional. It is not recommended to ingest rocks, minerals and crystals in any form!**)

We think this book is unique in that it combines an in-depth look at both the geological data about rocks and also the metaphysics of crystal healing. Presented within these pages is information that we hope will stimulate thought using both sides of the brain; the logical, analytical, left hemisphere, and the creative and intuitive right hemisphere. While there are many reliable sources of information on either topic, we found it difficult to find one good source that gives equal and proper attention to both realms. This book will try to focus on all aspects, presenting as balanced and comprehensive a story as we could tell. It is our feeling that as we have increased our own understanding of these diverse topics we have developed an even greater appreciation for the beauty of the rocks in our collection and the world from which they come. We hope this book will in part, kindle the same appreciation in our readers.

We tried our best to represent all pertinent aspects in this guide and we tried to be as accurate as possible, cross-referencing multiple sources to confirm data. Our intention is to provide a sound basis for exploration that will be useful to anybody with an interest in the topic. It is possible we have made errors or oversights and would welcome constructive input should you care to share your findings with us. We are always learning and we hope the adventure continues!

The photographs that accompany each listing in this book are of specimens from our personal, family collection.

STARTING A ROCK COLLECTION

Like so many people, we began collecting rocks as small children. Over the years we have collected rocks on vacations and travels, in shops and markets, backyards and parks. Our two sons have also grown-up with a similar fascination and passion for all things geologic. As their collection has grown, so has their knowledge about their specimens. Naturally, this leads to having desires for acquiring others to "fill in the holes", or to find a related mineral, or simply to seek out something they have seen or explored through research or from reading.

This is a very typical and organic way to build a collection. We are not geologists, so we are not motivated strictly by classification sorting or having a rigid, structured plan for organization. We do make a point to educate ourselves as thoroughly as possible about the chemical and crystal properties, mineral family, hardness, streak color and the habits, occurrences, uses and history of our specimens. We all feel this is worthwhile practical knowledge for both adults and children. We are driven by the passion for the beautiful array of rocks, minerals, crystals and gems available from Planet Earth.

No matter your budget or space, it is possible for anybody to gather his or her own rock collection. What we have learned is that a stone need not be costly to be precious, and there is no rock too small or ordinary to be desirable in some sense. We urge you to find rocks that please you whether simply for visual appearance, or to satisfy a desire no matter the reason. There is no "right" or "wrong" rock or crystal to collect. We hope that by reading this guide you are further inspired to continue your exploration into the exceptional wonders of rocks and to pass that enthusiasm on to others.

ORGANIZATION OF THIS GUIDEBOOK

There are a multitude of methods we could have used to organize this guidebook. In this second edition of the guidebook, certain minerals are grouped together by mineral families. The individual minerals in these groups still appear alphabetically in the Alphabetical Index of Specimens, but in some instances, appear in the body of the book under the name of their mineral group. We do this in the hope of presenting a better understanding of how specific and diverse minerals relate to each other.

Typically, a geology or mineralogy book could be divided by mineral family groups, crystal structure, hardness or streak color. All are valid methods and we have included indexes using these reference systems at the end of the book. For the healers and users of stones and crystals for meditation or metaphysical purposes, a book of this nature could be divided by chakra, by healing properties or astrological reference. We have also indexed our specimens, using these methods at the end of book.

For simplicity, we have decided to index this book in alphabetical order, making it quick and simple to find a particular rock or crystal. Our hope is that this book should be more than a basic reference guide. We wish it be a source of inspiration for learning about all facets of rocks and crystals; expanding knowledge and furthering the quest for answers. If your interest in collecting rocks has been strictly geologic in nature *(left-brain activity)*, we urge you to also explore and consider the references for the metaphysical properties of the specimens listed in this book. Likewise if your desire is to learn about the special properties of stones or how to use them as healing crystals *(right-brain activity)*, we strongly believe that some basic knowledge of their physical properties adds greatly to the experience of using crystals.

Please explore the categorized and cross-referenced indexes at the back of the book for further investigation. Enjoy!

ALPHABETICAL INDEX OF SPECIMENS

7

ALPHABETICAL INDEX OF SPECIMENS
CON'T

AEGERINE

FAMILY: Pyroxine Inosilicate

CHEMISTRY: $NaFe^{3+}[Si_2O_6]$

HARDNESS: 6

STREAK: White

SG: 3.50—3.60

STRUCTURE: Monoclinic Crystal

AEGERINE

AEGERINE is a sodium-rich clinopyroxine mineral found in certain alkali mineral deposits such as granite pegmatites. It occurs in diverse areas such as Greenland, Scotland, and the United States. Aegerine can be found in radiating, striated crystals or as small monolithic slab crystals, usually depending on the available growing space within the host cavity. Because it is a pyroxene mineral, its common color is green, however it is such a dark shade that it is often mistaken for black. Thinner crystals held up to strong light will clearly show a beautiful and slightly translucent dark green hue.

Aegerine is used as a personal spiritual energy booster and also for energizing a room or specific location. It is a high vibration crystal which can help one to approach aspects of the higher self. Used in group situations, aegerine can raise the energy level of a meeting or social gathering. Physically, aegerine is a booster for the immune system and is said to help eliminate toxins from the body's systems and organs. While not very common in the healing market, very nice specimens can sometimes be found at rock and mineral retailers. It is a fairly heavy mineral with great tactile presence, making it ideal for holding during meditation or rest.

ASTROLOGICAL: *Taurus,* **CHAKRA:** *Base,* **ELEMENT:** *Water*

AGATES

AGATE

FAMILY: Cryptocrystalline Silicate

CHEMISTRY: SiO_2

HARDNESS: 7

STREAK: White

SG: 2.58—2.64

STRUCTURE: Rhombohedral Crystal

AGATE SLICES

AGATE is a mineral group closely related to jasper, both being multi-colored and/or layered varieties of chalcedony. Agates are generally found filling cavities in igneous rock deposits *(often basalt)*, where silicate-rich fluids have deposited successive layers of microscopic quartz crystals *(cryptocrystalline silicate)*. Agate nodules can also be found where surrounding rock has eroded away. Varying amounts of silica and mineral impurities are responsible for the almost endless variety of colored banding that are typical of agates. Unlike jaspers, agates are normally semi-translucent and this is a good first step in differentiating the two.

Sliced and polished agate specimens are a common and attractive method for displaying the colorful and layered features found in the interior structure of agates. However, the vast majority of agate slices, book-ends and other decor items are artificially dyed to enhance color or to create colors not found in natural agate. A reputable dealer can show you which of the beautifully colored agates are authentic in color. Agates have been accessible and useful as decorative material for many centuries in a wide range of locations throughout the world. A very wide variety of agates are still available commercially, and can constitute a significant portion of many collections, including our own.

Agates are useful as healing crystals, and have properties specific to each type. These properties are discussed in detail in the individual listings on the following pages. One feature that all agates share is their affinity to plants and all things that grow, making them excellent for gardens, houseplants and farms.

BLUE LACE AGATE

BLUE LACE AGATE is a delicate mix of semi-translucent, pale-blue hues and semi-opaque white striations. The finest specimens have long come from mines in South Africa, however that source is reported to be nearly depleted. Several other minor sources that produce blue lace agate include Brazil, China, India and the United States.

Blue lace agate is calming and tranquilizing. It encourages verbal expression and heart-felt communication. It can neutralize anger and helps to elevate energy and dispel bad moods. Blue lace agate heals and repairs bones, helps with skin problems, soothes red eyes, inflammation, infection and fever.

ASTROLOGICAL: *Pisces,* **CHAKRA:** *Throat,* **ELEMENT:** *Water and Air*

BOTSWANA AGATE

PINK BOTSWANA AGATE

BOTSWANA AGATE has characteristic brown and tan banding, alternating with white chalcedony. Botswana, in southern Africa is the only known source for this particular agate. Botswana is better known mineralogically for its mining of gold, copper, diamonds and uranium but is also a good source for many minor semi-precious stones such as these.

Botswana agate stimulates personal exploration and encourages one to seek out new paths to enlightenment and higher states. It enhances energy, yet discourages impulsiveness. It improves memory and sharpens mental abilities. It can act as both a stimulant and an anti-depressant.

ASTROLOGICAL: *Scorpio,* **CHAKRA:** *Sacral,* **ELEMENT:** *Air*

PINK BOTSWANA AGATE has very similar features to Botswana agate but the colors are typically very pale pinks, peaches and white instead of shades of brown and white. Pink Botswana agate displays far more translucence that the brown-hued types. Botswana, in southern Africa is the only known source for these particular agates.

Pink Botswana agate releases repressed emotions and feelings. It cleanses the aura and is of great use during meditation. It helps the body assimilate oxygen, which is good for the skin, the brain and circulatory system. It also helps to encourage recovery from chronic depression or sadness.

ASTROLOGICAL: *Gemini,* **CHAKRA:** *Crown,* **ELEMENT:** *Air and Fire*

CRAZY LACE AGATE

FIRE AGATE

CRAZY LACE AGATE occurs in Mexico and is sometimes similar in pattern to blue lace agate, however crazy lace agate tends to be formed with shades of brown, orange or yellow banding alternating with white chalcedony. Some of the better examples have extremely rich and vivid swirls of color.

Crazy lace agate can help heal emotional pain and it is believed to clear away bad energies that block personal well-being and growth. It sharpens the mind and aids with concentration. Crazy lace agate stimulates the entire body and balances the function of the body's organs.

ASTROLOGICAL: *Gemini,* **CHAKRA:** *Solar Plexus,* **ELEMENT:** *Earth*

FIRE AGATE usually has a deep, rich reddish orange hue with occasional dark inclusions, giving it the appearance of a glowing ember, sometimes stunningly so! It's slight translucence highlights this effect even more. Typically this agate comes from the Southwestern United States and some of the finer specimens display a deep opalescence, making it suitable for jewelry purposes. It is fairly uncommon, and we are lucky to have a few small but attractive specimens in our family collection.

Fire agate can give guidance towards taking action, making decisions, and good choices. It can release passionate emotions, and unknown desires and reveal the path towards achieving new goals. Physically, fire agate treats glands and the endocrine system, the intestines and colon, and regulates metabolism. It also improves poor night vision.

ASTROLOGICAL: *Gemini,* **CHAKRA:** *Sacral,* **ELEMENT:** *Fire*

MOSS AGATE

TREE AGATE

MOSS AGATE has an appearance of dendritic moss or a blue-cheese pattern although its color is not from organic nor vegetative matter. The green colorings are usually from oxides of manganese, chrome or iron. Moss agate displays areas of translucent quartz or clear agate. Moss agate is one of the few stones from the chalcedony family that is classified as an agate without having any requisite banding or swirl patterns.

Moss agate soothes the soul and reveals beauty in all surroundings. It eases discomfort due to bad weather and comforts children during storms. Protects against environmental pollutants and EM radiation. Speeds recovery from colds, and acts as an anti-inflammatory for the sinuses. It detoxifies the circulatory and digestive systems, and boosts the immune system. Moss agate is frequently used by healers to loosen stiffness of joints, especially in the fingers and wrists.

ASTROLOGICAL: *Gemini,* **CHAKRA:** *Crown,* **ELEMENT:** *Earth*

TREE AGATE is predominantly white, with shades of green, dendritic patterns sparsely covering the stone. While similar in general appearance to moss agate, the opposite would be true; moss agate is predominantly green with sparse white patterning.

Tree agate is believed to provide a powerful connection to Mother Nature. It is a good stone for healing or maintaining plants and trees, for enjoying the outdoors and to use during meditation outdoors. Helps limit the discomfort of neuralgia. It is a highly supportive stone and promotes a healthy attitude. It can boost the immune system and is said to regulate the balance of water and fluids in the body.

ASTROLOGICAL: *Gemini,* **CHAKRA:** *Crown,* **ELEMENT:** *Earth*

ORANGE SNAKESKIN AGATE

WHITE SNAKESKIN AGATE

SNAKESKIN AGATE comes in several basic color schemes; white and orange-ish brown being most common. It forms as nodules or small lumps and has a course texture pattern, similar to the skin of a snake or a lizard. It takes a gentle polish very well but cutting or cabbing may remove some of the most desirable surface textures. Idaho is the main source for this agate. Some silicified* corals also bear the descriptive name snakeskin agate, but it most probable that the varieties found in Idaho are actual textured varieties of chalcedony and not fossiliferous in origin.

Snakeskin agate brings out the cheerful side of one's personality and assists in easing worries. A good and trustworthy stone, it is helpful for finding lost items. Treats stomach disorders and indigestion, smooths skin and wrinkles. It counteracts the toxins in snakebites.

ASTROLOGICAL: *Aries,* **CHAKRA:** *Sacral,* **ELEMENT:** *Air*

(silicification)—Some minerals or organic materials undergo a process called pseudomorphism, which means that their original chemical structure has been replaced by another mineral. When matter is replaced by quartz or other silicate it is said to have "silicified".

SULEMANI AGATE

SULEMANI AGATE, aka Solemani Agate or Solemani Stone, the name of this agate derives from India where the original name is Sulemani Hakik (hakik meaning quartz). This agate displays banding patterns very similar to Botswana agate or Sardonyx, and similarly, tends to occur mainly in hues of black, white, and dark brown. Technically, besides location of origin, Sulemani, Botswana, and Sardonyx, could be considered to be the same variety of agate.

Sulemani agate not only provides the bearer with great energy and strength, but has also been used traditionally to ward off evil, negative energy, and to protect against harm.

ASTROLOGICAL: *Sagittarius,* **CHAKRA:** *Sacral,* **ELEMENT:** *Water*

Related entries in this book: Botswana Agate, Sardonyx

TURRITELLA AGATE

TURRITELLA AGATE is unusual in its appearance and its origin. Technically, it is not an agate at all, that term being specific to banded varieties of chalcedony, which form as layered deposits of cryptocrystalline silicates. Turritella agate is a fossiliferous stone, being formed by silicification* of fresh water snail shells called Elimia Tenera that lived approximately 50 Million years ago in what is now Montana. Oddly, these sea shells are named for a closely related, but separate species of sea water snail shells called Turritella Communis. This misnomer is probably due to an early misclassification.

Turritella agate opens channels of communication between ourselves and the plant and animal species of our world. It is in tune with information for healing the planet. These energies can be focused on a specific location by placing the stone on a map or photo of endangered place on the planet or where flora or fauna are in danger of extinction. It is also of great use to help find lost items. Reduces swelling and stiffness in the hands and feet. Eases problems related to aging and helps with digestion, improving conditions like Crohn's disease or gastroenteritis. Turritella is a potent healer of skin rashes and can soothe itches from insect bites.

NOTE: *Unlike most other agates, turritella agate has a trigonal hexagon crystal structure, not a monoclinic crystal structure.*

ASTROLOGICAL: *Gemini,* **CHAKRA:** *Base,* **ELEMENT:** *Water*

(silicification)—Some minerals or organic materials undergo a process called pseudomorphism, which means that their original chemical structure has been replaced by another mineral. When matter is replaced by quartz or other silicate it is said to have "silicified".

WHITE AGATE

WHITE AGATE differs from white chalcedony only in the respect that agates are banded, and chalcedony is typically smooth and uniform in color. The bands on white agate are sometimes creamy, or slightly tinged with yellow or brown. The lightest areas are highly translucent. Specimens in our family collection that are as much as 1 inch thick, display a beautiful inner-glow when held to light. One of the first written, historical descriptions of agate use is in ancient Greek texts describing white agate. To this day, it is sometimes referred to as "Greek stone" even though most modern examples come from Brazil and elsewhere in South and North America.

   White agate is also known as the peace stone, and can bring serenity and harmony by holding or gazing at it while meditating. It can be helpful for relaxation and focus, and clarity of mind and spirit. White agate assists with breaking bad habits or addiction. When used with ruby or with fuchsite it is an excellent aid to help quit smoking.

ASTROLOGICAL: *None,* **CHAKRA:** *Crown,* **ELEMENT:** *Earth*

YELLOW AGATE

YELLOW AGATE is one of the less common forms of agates and is frequently, falsely, sold as white or gray agate that has been dyed. Dyed yellow agate is often easy to distinguish in that it takes high saturation levels of yellow dye to turn a light-colored agate to yellow, so it may not be as translucent as a true yellow agate should be.

Yellow agate activates inner strength and courage and is a reminder to seek-out and honor truth. It promotes good manners in children, brings happiness, intelligence, prosperity and longevity. Yellow agate carvings can be used as totems of fertility. It helps find and keep new friends. It is helpful to farmers and those interested in growing their own plants. Yellow agate stimulates the intake of oxygen for the entire body, and protects cells from the ill-effects of x-rays. It is useful also, for relieving upset stomachs and treating lung ailments. It can help to ease depression and remove lethargy, and treats conditions of the throat and skin.

ASTROLOGICAL: *Virgo,* **CHAKRA:** *Heart,* **ELEMENT:** *Earth*

AMAZONITE

FAMILY: Tectosilicates

CHEMISTRY: $KAlSi_3O_8$

HARDNESS: 6–6.5

STREAK: White

SG: 2.58–2.59

STRUCTURE: Triclinic Crystal

AMAZONITE

AMAZONITE is the blue-green gemstone variety of microcline, an orthoclase feldspar. It is supposedly named after the color of the Amazon river, however no deposits of it have ever been found there. Amazonite is currently mined and exported to the market from Madagascar, Brazil, Russia and Colorado and Virginia in the United States. Some finer specimens can be mistaken for jade because if its rich green color and pearlescent sheen. Its color comes mostly from impurities of lead, and it cleaves or fractures easily, so care must be taken if cutting or cabbing for jewelry purposes. It is usually hard enough to tumble and polish successfully, however it is also commonly sold in its raw form, which tends to display its fine texture and delicate shine faithfully.

Amazonite is a stone of empowerment, but at the same time it helps with maintaining personal integrity and sense of fairness. It opens the bearer to all forms of communication, including clairvoyance, and vision. Physically, healers use amazonite to prevent tooth decay and osteoporosis by increasing absorption of calcium. It promotes health of the liver, throat, thyroid, the nervous system and brain. It also helps to relieve pinched nerves, spinal afflictions, arthritis, and bursitis.

ASTROLOGICAL: *Virgo,* **CHAKRA:** *Throat,* **ELEMENT:** *Earth*

Related entry in this book: Orthoclase feldspar

AMBER

FAMILY: Mineraloid

CHEMISTRY: Fossiliferous

HARDNESS: 2.0–2.5

STREAK: White

SG: 1.0–1.1

STRUCTURE: Organic Crystal

AMBER

AMBER is a beautifully translucent, fossilized remnant of tree resin *(not sap)* and dates from between 125 million and 350 million years old. It has been used for decoration and ornament since very early in human history. Amber is also famous for sometimes having perfectly preserved inclusions of ancient insect and plant life. Specimens such as these, are important to both collectors and scientists since they offer a sample of living material unchanged by decomposition and weathering. Jewelry and collector's pieces of amber remain as popular today as they have been for millennia. In recent years, several new sites have been discovered in South America, bringing even more precious amber to market. A simultaneous increase in copal imports to America has led to some confusion and some unreliable *(sometimes unscrupulous)* marketing. To be clear, amber is truly fossilized and is considered a valuable, semi-precious mineraloid. Copal is neither of those but can be attractive and used as ornamentation nonetheless.

Amber is wonderfully light and glowing and is used for cleansing the chakras and maintaining the mind and body balance. Historically, amber has been used as pain-reliever when used in meditation, and it brings good-fortune to mothers and their new-born children.

ASTROLOGICAL: *Taurus*, **CHAKRA:** *Crown*, **ELEMENT:** *All*

Related entries in this book: Copal ,Jet

RED AMBER

AMETHYST

AMETHYST

FAMILY: Quartz Silicates

CHEMISTRY: SiO_2 (with traces of Fe^{3+})

HARDNESS: 7

STREAK: White

SG: 2.58–2.59

STRUCTURE: Trigonal Hexagon Crystal

AMETHYST

AMETHYST is a variety of crystalline silicate, identical in chemistry and structure to quartz, and citrine. Its purple coloration is due to trace impurities of iron and other naturally irradiated minerals. Its transparent purple color, attractive crystals and hardness make it a very popular semi-precious gemstone. Amethyst has been used as jewelry for millennia and has always been popular with royalty in societies where purple is considered a regal color. Amethyst typically forms in small cavities and geodes where precipitation of silica-rich fluids can collect and grow crystals undisturbed. Some of the finest crystals come from Brazil where enormous amethyst bearing geodes are large enough to walk in. Other quality amethyst comes from Siberia, Sri Lanka, Uruguay, and the United States, however Zambia is currently the largest exporter.

Amethyst is a stone of great protection. It relieves stress, promotes relaxation and brings awareness and intuition. It cleanses the spirit and body and opens the chakras. It can help bring deep and tranquil sleep. Amethyst strengthens the immune system, and is a general healer that balances the metabolism, fights the ills of cancer, relieves pain, swelling, bruising, headaches, hearing and respiratory ailments.

ASTROLOGICAL: *Virgo*, **CHAKRA:** *Third Eye*, **ELEMENT:** *Air*

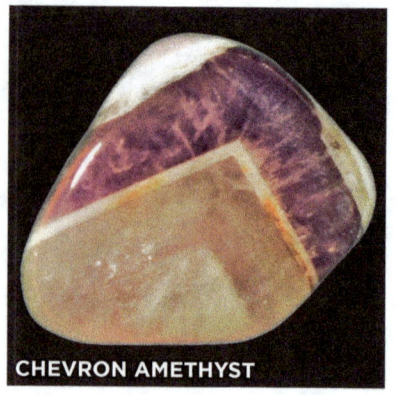

CHEVRON AMETHYST

CHEVRON PRASIOLITE

CHEVRON AMETHYST is formed with bands of clear quartz, milky quartz or white chalcedony. It is also sometimes known as banded amethyst. This banding occurs when the chemical or ion composition changes during the formation of the crystal.

Chevron amethyst dispels resistance to change and repels negative energy of all kinds. It creates a strong healing field around the user. It helps to overcome addictions and destructive habits. Chevron amethyst calms or stimulates the mind, whichever is appropriate at the time, meaning that it can assist in treating both lethargy or insomnia.

ASTROLOGICAL: *Aquarius,* **CHAKRA:** *Sacral,* **ELEMENT:** *Air*

PRASIOLITE is green amethyst. It is nowhere near as common in nature as purple amethyst or golden citrine, and is similarly scarce in mineral shops. However it is not too costly and makes very attractive tumble-polished stones, which is the most likely form you will find specimens of prasiolite. Healers suggest prasiolite as a detoxifying crystal for both the mind and body, and it used as an amulet or talisman for improved general health. It rings clarity, focus and enhanced intuition.

ASTROLOGICAL: *Virgo,* **CHAKRA:** *Third Eye,* **ELEMENT:** *Air*

Related entries in this book: Ametrine, Citrine, Quartz

AMETRINE

FAMILY: Quartz Silicates

CHEMISTRY: SiO_2 (with traces of Fe^{3+})

HARDNESS: 7

STREAK: White

SG: 2.58–2.59

STRUCTURE: Trigonal Hexagon Crystal

AMETRINE

AMETRINE is a naturally occurring crystalline silicate variety of quartz. It is a mixture of amethyst and citrine with zone's of purple and golden-yellow or light orange. Currently, almost all commercially available ametrine originates from deposits in Bolivia, although there are also some small mines in operation in Brazil and India. The distinct color zones within ametrine are due to differing oxidation states or irradiation variations of the iron impurities within the silicate available as the crystal forms. Much like amethyst and citrine, ametrine can form in conjunction with quartz and chalcedony, sometimes in the same pockets or geodes. Ametrine is not nearly as common or as readily available as amethyst or citrine, however quality specimens with distinct color-zoning are quite precious and desirable for jewelry purposes.

Metaphysical properties of ametrine correspond to the properties of both amethyst and citrine which can be found under those listings in this guide.

ASTROLOGICAL: *Virgo,* **CHAKRA:** *Third Eye,* **ELEMENT:** *Air*

Related entries in this book: Amethyst, Citrine, Quartz

ANDALUSITE (Chiastolite)

FAMILY: Nesosilicates

CHEMISTRY: Al_2SiO_5

HARDNESS: 6.5-7 .5

STREAK: White

SG: 3.17-3.19

STRUCTURE: Orthorhombic Crystal

ANDALUSITE

ANDALUSITE (CHIASTOLITE) is an aluminum nesosilicate mineral most commonly found in metamorphic rock formations. It is identical to kyanite in chemistry but with a different crystal structure. These polymorphs form under different temperature and pressure situations making the occurrence of kyanite or andalusite a handy reference with which mineralogists can infer conditions under which the metamorphic rocks formed. Chiastolite *(pictured here)* is a variety of andalusite whose crystals display distinctive cross-shaped black inclusions of graphite. It only rarely forms as clear crystals which are suitable for faceting and are sought after for jewelry purposes.

Chiastolite is a stone of harmony, security and protection from evil spirits and negative energy. It brings an awareness of the flow of life and how to find balance between personal needs and the needs of others. Chiastolite helps the bearer to more easily accept the signs of aging and fully experience the joys of the maturing process. It is a strong stone to use for the health of teeth and bones. It is used by healers to balance the blood and circulatory system as well as being a valuable aid for body to repair chromosome and cellular damage.

ASTROLOGICAL: *Cancer,* **CHAKRA:** *Sacral,* **ELEMENT***: Earth*

Related entry in this book: Kyanite

ANHYDRITE (Angelite)

FAMILY: Quartz Silicates

CHEMISTRY: $CaSO_4$

HARDNESS: 7

STREAK: White

SG: 2.80–2.90

STRUCTURE: Orthorhombic Crystal

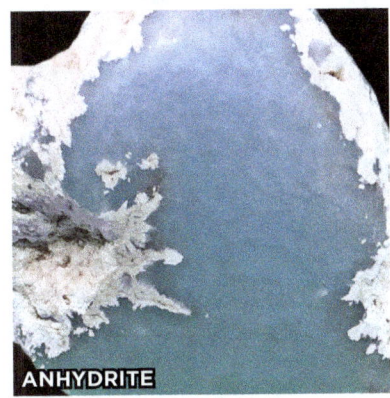

ANHYDRITE (ANGELITE) is an evaporate mineral most frequently found in deposits with gypsum. It has identical chemistry to gypsum and forms when gypsum dries out. It most commonly occurs in thick, massive deposits and only rarely is found in crystalline form, since prolonged contact with water or exposure to moisture makes anhydrite turn back into gypsum. In fact, the word anhydrite means "without water". Because of its reaction with water, most anhydrite deposits are only found deep within thick masses of gypsum, where moisture cannot reach it. Anhydrite can occasionally be found with a beautiful pale-blue tint, and has earned the popular name angelite due to its celestial color. It is not nearly as common in collections as gypsum *(e.g. desert roses)* and must be kept dry or in a sealed container to retain its anhydrous chemistry and prevent it from reverting to basic gypsum. Tumble-polished angelite stones are not so susceptible to this chemical metamorphism.

Angelite is used to raise one's state of awareness. It represents peace and assists in connecting with spirit guides, especially for the first time. It opens one's mind to finding and understanding the inner-self. Physically, it is helpful for treating conditions of the throat and reduces inflammation anywhere in the body. Angelite can help balance thyroid and metabolic functions, as well as repairing damaged tissue and improving the health of the circulatory system. In the crystal healing realm, the mineral celestite *(celestine)* is also often referred to as angelite. Mineralogically and chemically however, angelite and celestite, are completely unrelated minerals. See the listing for celestite in the book for more information.

ASTROLOGICAL: *Aquarius,* **CHAKRA:** *Throat,* **ELEMENT:** *Water*

Related entry in this book: Gypsum Desert Rose

ANYOLITE

FAMILY: Metamorphic Rock

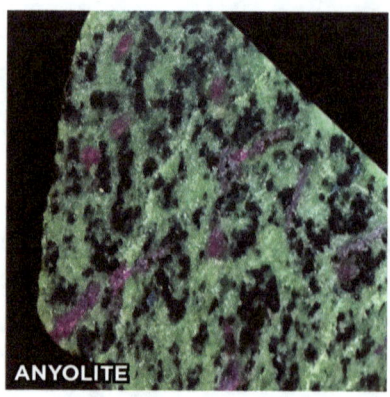

ANYOLITE

ANYOLITE is a dark green metamorphic rock known as zoisite, intergrown with red ruby crystals and small dark crystals of pargasite. Anyolite is also commonly called ruby in zoisite. Pink thulite and violet blue tanzanite are both other forms of zoisite but are not found in association with ruby. Because of its striking contrast and pattern, anyolite is mostly used for carvings and decorative objects, rather than for jewelry purposes. Anyolite was discovered in the 20th century in Tanzania which is still the only mining location for this beautiful stone.

Because of extensive worldwide distribution, anyolite has become a popular healing crystal. Common usages focus on empathy, spirituality and making connections between one's mind and heart. Meditational focus is on self-esteem and accessing repressed memories.

ASTROLOGICAL: *Aries*, **CHAKRA:** *Third Eye*, **ELEMENT:** *Earth*

Related entries in this book: Ruby, Thulite, Zoisite

APACHE TEARS

FAMILY: Mineraloids

CHEMISTRY: SiO_2 + MgO,Fe_3O_4

HARDNESS: 5.5

STREAK: White

SG: 2.40-2.60

STRUCTURE: Amorphous Rock (noncrystalline)

APACHE TEARS NODULE

APACHE TEARS are small nodules of an igneous volcanic glass that cooled quickly in the presence of water. After deposition, the obsidian hydrated, and parts of its structure turned to white perlite, which is an identifying feature of true apache tears. Upon weathering of the surrounding bedding area, apache tears being harder than the surrounding rock, may be exposed. Apache Tears are most commonly black in color, but can also occur in dark shades of brown and gray and occasionally dark red. Obsidian varieties all share similar chemistry, with silica being the main constituent, and various ratios of other metals or oxides being responsible for variations in color and opacity.

Like all obsidian, apache tears are technically considered a glass, not a rock. For most mineralogical classification purposes all obsidian are placed in a general category called mineraloids. Glasses *(obsidian, Apache tears, fulgarite, goldstone)*, organics *(amber, copal, jet)* and concretions *(moqui marble, boji stone, septarian)* are all members of the mineraloid family. Information on these mineraloids can be found under their individual listings in this book.

The name "Apache tear" came from the story of a devastating defeat against the U.S. cavalry in the 1870's, in which all of the Apache warriors were killed. Legend says that the wives and families of these warriors cried tears that hardened into stone as they hit the ground.

Apache tears give support to the heart and the soul, protecting the bearer from sorrow. Apache tears provide protection from negativity and emotional pain. Because of their ability to reach deep within the spirit as a healing crystal, Apache tears should be cleansed regularly in sunshine or cool water.

ASTROLOGICAL: *Aries,* **CHAKRA:** *Root,* **ELEMENT:** *Earth*

Related entry in this book: Obsidian

APATITE

FAMILY: Phosphates

CHEMISTRY: Ca5(PO4)3(F,Cl,OH)

HARDNESS: 5 (Defining)

STREAK: White

SG: 3.17–3.23

STRUCTURE: Hexagonal Crystal

BLUE APATITE

APATITE is the name of both a mineral group and a generic term for many phosphates. Gem quality apatite is found in a variety of colors due to the presence of different rare earth elements or from exposure natural irradiation. Because of its wide color range, apatite has often been mistaken for many other gemstones, including topaz, tourmaline, and various kinds of beryl. In fact, its name derives from the greek word "apate", meaning deceit. Apatite is an extremely common mineral and is the main source of phosphorous in most soils around the world. Major sources of gem quality apatite are Brazil, Burma and Mexico, with lesser quality varieties being abundant throughout the world. Moon rocks brought back to Earth during the Apollo space program contained significant traces of apatite mineral. Non-gem apatite is a soft, flaky rock that while unsuitable for cutting, can take a light polish fairly well. We have had great success forming and smoothing green and blue apatite at home with grinders and polishing cloths, producing some beautiful palm stones and decorative shapes. Apatite is abundant and relatively inexpensive at collector or healing crystal shops.

BLUE APATITE inspires the soul and brings self-confidence and clear thought to the bearer. It helps with nail problems, allergies, arthritis, bones and muscles. Apatite is generally beneficial to the brain and the entire nervous system.

ASTROLOGICAL: *Gemini*, **CHAKRA:** *Throat*, **ELEMENT:** *Air*

GREEN APATITE MASS

YELLOW APATITE IN MATRIX

GREEN APATITE clears confusion, apathy or negativity, and stimulates the bearer to expand the intellect and seek truth. It is a reliable ally for personal growth. Physically, it improves hand-eye coordination, especially in children. Green apatite protects pets and wildlife from harm.

ASTROLOGICAL: *Gemini,* **CHAKRA:** *Solar Plexus,* **ELEMENT:** *Air*

YELLOW APATITE releases pent-up anger and frustration. It brings lucid dreams and assists with astral travel and also reminds us to be involved in humanitarian deeds. It rids the body of toxins, treats depression and lethargy and improves concentration, thus promoting good study habits, especially in children. Yellow apatite improves poor digestion and can help remove cellulite.

ASTROLOGICAL: *Gemini,* **CHAKRA:** *Throat,* **ELEMENT:** *Air*

APOPHYLLITE

APOPHYLLITE, STILBITE & CALCITE CLUSTER

RED APOPHYLLITE

APOPHYLLITE CLUSTER

APOPHYLLITE

FAMILY: Phylosilicate Zeolites

CHEMISTRY: $(K,Na)Ca_4Si_8O_2(F,OH)\ 8H_2O$

HARDNESS: 4.5–5

STREAK: White

SG: 2.32–2.41

STRUCTURE: Tetragonal Crystal

APOPHYLLITE

APOPHYLLITE is very popular with collectors of all types because of its beautiful and extraordinary pyramidal crystal shape. This feature, in addition to its beautiful clarity make it a highly sought-after gem for collectors and healers. It is occasionally found in pale shades of pink and green in addition to its most common, colorless form. Crystals of high quality, display prismatic effects, showing rainbow refractions. The name apophyllite comes from the ancient Greek word for "leaf" or "flake", which describes the cleavage tendencies of apophyllite. While not a zeolite mineral itself, it occurs most frequently with zeolites such as stilbite and heulandite. Very commonly, pyramidal crystals are sold individually, although they tend to grow in conglomerate masses with multiple prism points. Apophyllite can be found in Mexico and parts of Europe but the major source for quality crystals is the Deccan Traps region in India. Imports have increased dramatically the past few years and larger, display-quality clusters are frequently found at rock shops and online from many reputable dealers. We have found several stunning apophyllite clusters at stores for very reasonable prices and they are excellent and admired additions to the family collection.

Apophyllite clears and prepares the mind for meditation and helps attune psychic perception. Apophyllite is a stone of blissful inner peace. It brings clarity of thought when seeking visions so it is useful to reiki healers, meditation guides and energy workers. Helps repair damaged DNA, improves memory and retention of dreams. Clear apophyllite treats the urinary system and improves poor bladder function. It is also recommended by healers to treat the colon, mucous membranes and the respiratory system.

ASTROLOGICAL: *Libra,* **CHAKRA:** *Third Eye,* **ELEMENT:** *Earth*

AQUAMARINE

FAMILY: Cyclosilicates

CHEMISTRY: $Be_3 Al_2 (Si_6 O_{18})$

HARDNESS: 7.5–8

STREAK: White

SG: 2.84–2.91

STRUCTURE: Hexagonal Crystal

AQUAMARINE

AQUAMARINE is the blue or blue-green variety of the famous beryl family of minerals. Its name comes from the Latin, "aqua marina", meaning sea blue. Pure beryl is colorless (called goshenite), but when tinted by varying ratios of two iron ions *(Fe$_2$+ and Fe$_3$+)*, beryl turns to extremely rich shades of blue *(blue beryl)*, green *(emerald)*, yellow *(heliodor)*, red *(red emerald)* as well as aquamarine. Like many gemstones, aquamarine is found mainly in pegmatite* formations. Madagascar, Tanzania and Colorado and Montana in the United States are all major producers of aquamarine and other beryls. Low-grade, rough and unpolished specimens are readily available and inexpensive but gem-quality specimens are faceted for the jewelry market and are quite precious. Aquamarine always appears naturally with greenish tint, but after heat-treating and cutting, blue becomes more apparent. Strong sunlight can fade the blue hues, and in fact all beryls will fade and eventually turn colorless if exposed to too much sunlight. Gem quality aquamarine has almost always been heat-treated.

Aquamarine is a stone of strength and courage. It also has calming energies that reduce stress and quiet the mind. Aquamarine has a special bond with sensitive people. It promotes acceptance and tolerance and lends support to those overwhelmed by responsibility. It sharpens the senses and the mind and clears confusion. A very useful stone for closure on all levels, aquamarine is a comfort stone which increases one's confidence. Healers use aquamarine to strengthen the immune system and alleviate allergies such as hay fever and rashes. It is also used to treat sore throats and swollen glands.

ASTROLOGICAL: *Aries,* **CHAKRA:** *Throat,* **ELEMENT:** *Water*

Related entries in this book: Emerald, Goshenite , Heliuodor, Morganite

*(pegmatite)—The last parts of an igneous intrusion to cool. Because pegmatites are rich in heavier and/or metallic minerals, they are the main source of large, colorful crystals and precious gemstones on the Earth's crust.

ARAGONITE

FAMILY: Carbonates

CHEMISTRY: $CaCO_3$

HARDNESS: 3.5–4

STREAK: White

SG: 2.93–2.94

STRUCTURE: Orthorhombic Crystal

ORANGE ARAGONITE

ARAGONITE is a calcium carbonate mineral, related to calcite by identical chemistry. The two frequently occur together and can be distinguished by their different crystal habits. Calcite has an orthorhombic crystal structure and tends to form in masses, whereas aragonite crystals are trigonal and often form dramatic spiked, branched or hexagonal crystal groups. It takes its name from the Aragon region in Spain where it was first identified.

Aragonite centers and grounds physical energies and is very useful in time of stress. It is a great companion crystal for meditation and spiritual development. Aragonite reveals insight into the causes of problems, aids concentration and brings tolerance and flexibility to the mind. Physical uses include warming and soothing the extremities. It combats vitamin deficiencies and is generally useful for combating disease. Aragonite can help to stop twitches and muscle spasms. It also helps with general aches and pains.

Related entry in this book: Calcite

WHITE ARAGONITE

ORANGE ARAGONITE forms mostly twin and trilling crystal clusters with well-defined hexagonal crystal faces and is beautifully translucent. The most abundant sources of well-formed aragonite clusters available today are Mexico, Morocco and New Mexico and Arizona in the United States. Orange aragonite has become a common find at better mineral and rock retailers. These beautiful clusters make excellent display pieces and are definitely some of the author's favorite specimens.

WHITE ARAGONITE clusters, displaying beautiful sprays of prism crystals are traditionally referred to as "floss flori", from the Latin for, "flower of iron" because aragonite clusters may be commonly found associated with iron ore deposits.

ASTROLOGICAL: *Gemini,* **CHAKRA:** *Throat,* **ELEMENT:** *Air*

Related entry in this book: Calcite

ARCANITE

FAMILY: Sulfates

CHEMISTRY: K_2SO_4

HARDNESS: 2

STREAK: White

SG: 2.66

STRUCTURE: Orthorhombic

ARCANITE

ARCANITE is a sulfate mineral that was discovered in Orange County, California in the mid-19th century and has more recently been found in Peru, Italy and Australia. However these occurrences are all massive or granular forms of the mineral, and the only attractive crystals of arcanite are lab-grown for the collector market. These crystals can reach substantial sizes up to 2 or 3 inches, with beautiful pyramidal forms, double termination and rich coloration. Root beer is the most common color but orange, red and many other hues have been produced. For their size, the crystals are surprisingly affordable and can be easily found at rock shops and online sellers.

Being manufactured and not natural, there is little information about arcanite being used for crystal healing purposes. It is fairly new to the market, so perhaps the spiritual community will discover unique uses for this attractive mineral.

ASTROLOGICAL: *None Known,* **CHAKRA:** *None Known,* **ELEMENT:** *None Known*

ASTROPHYLLITE

FAMILY: Inosilicates Zeolites

CHEMISTRY: $(K,Na)_3 (Fe++,Mn)_7$

HARDNESS: 3-4

STREAK: Yellow to Brown

SG: 3.20–3.40

STRUCTURE: Triclinic Crystal

ASTROPHYLLITE is an uncommon inosilicate mineral usually found in igneous rock formations and associated with feldspar, mica, titanite, zircon and several less common substances. Its typical crystal habit is bladed or stellate *(star-shaped)*. Its name derives from two Greek words, "astro", meaning star, and "phyllon", meaning leaf. Due to its remarkable shine, depth and star patterned crystals it is quite popular as an ornamental stone. It is typically cut in large cabochons for pendants or polished into palm stones for use, since these larger sizes are best suited to display the fine patterning and light-scattering qualities of this stone. Locations where astrophyllite is currently being mined are Canada, Norway, Mongolia, Russia and Greenland. It is not very common in mineral collector and healing crystal shops so it remains fairly costly. Unpolished specimens are extremely difficult to find as most astrophyllite goes to the jewelry and ornament industries. We are lucky to have one sizable astrophyllite specimen in our collection and it displays excellent shimmer, shine and chatoyance *(cat's eye effect)*. The metallic, silver colored stellate crystals contrast brightly with the generally orange-brown felsic *(feldspar and silicate)* base and mica, titanite, zircon inclusions.

Astrophyllite is a stone of self-awareness and acceptance. It promotes personal well-being and balance to the bearer. With this comes clearer vision of direction and desire for growth and change, so it is an excellent ally for releasing unhealthy patterns and habits. Astrophyllite can be used for help for people dealing with ADD, ADHD and anxiety. It helps with weight loss and removing fat deposits. Also believed to promote cellular regeneration.

ASTROLOGICAL: *Virgo,* **CHAKRA:** *All,* **ELEMENT:** *Air*

AVENTURINE

FAMILY: Cryptocrystalline Silicates

CHEMISTRY: SiO_2

HARDNESS: 6-7

STREAK: White

SG: 2.64-2.69

STRUCTURE: Trigonal Hexagon Crystal

PINK AVENTURINE

GREEN AVENTURINE

AVENTURINE is a common form of chalcedony *(quartz silicate)*. Unlike agates, it is not translucent and it is not banded like jaspers. This particular chalcedony mineral has unique type of multi-level shine referred to as "aventurescence". This effect is due to sparkling inclusions of chromium-rich fuchsite mica. There is common misidentification of this mineral since sunstone and other colorful feldspar varieties are sometimes referred to as aventurine. Spain and Russia are both substantial sources of many varieties of aventurine, but India is the main foreign producer. Quarries of green aventurine are found in Vermont, Wisconsin and Virginia making it abundant enough in the United States that it is commonly used as a landscape stone.

Aventurine is a stone of prosperity and is purported to bring wealth and success to the bearer. It reinforces leadership qualities and decisiveness. It promotes compassion and empathy towards all that surrounds you. It stabilizes one's state of mind, stimulates perception and enhances creativity, and aids in seeing alternatives and new possibilities. Aventurine tempers anger and irritation, balances blood pressure, stimulates the metabolism and lowers cholesterol. Healers promote its anti-inflammatory effect, and use it for treating allergies, migraines, and soothing the eyes. It helps heal the lungs, sinuses, heart and muscular systems.

ASTROLOGICAL: *Aries,* **CHAKRA:** *Heart,* **ELEMENT:** *Air*

PINK AVENTURINE (Unpolished)

GREEN AVENTURINE (Unpolished)

AXINITE

FAMILY: Sorosilicates

CHEMISTRY: $(Ca,Fe,Mn)_3Al_2BO_3Si_4O_{12}OH$

HARDNESS: 6-7

STREAK: White

SG: 3.31–3.34

STRUCTURE: Triclinic Crystal

AXINITE

AXINITE is actually a series of sorosilicate minerals named for their blade-like crystals. Because it is a triclinic crystal, it displays no visible symmetry but has randomly oriented sprays of bright colors and metallic streaks corresponding to the inclusion of several different metals. Iron, manganese, and magnesium appear in varying degrees of saturation within the stone, determining its specific name. The specimen from our family collection that is pictured here, is most likely a tinzinite crystal, composed primarily of silica with inclusions of iron and manganese. Varieties of axinite have been found in the United States in New Jersey and California mines, and around the world in such places as Brazil, France and Switzerland. Being a silicate mineral, it is very hard and takes well to cutting or polishing. While not an extremely common stone, it has no real commercial uses other than as collector's specimens or jewelry, so it can be found without too much effort; rarely as a crystal cluster, more often as tumble-polished stones. Fine specimens display a slight chatoyance, similar to the play of light as in tiger's eye crystals.

Axinite can help with acceptance of change without usual resistance, and provides a grounding force in all endeavors. Axinite aligns the body's energy, meridians and chakras and places the bearer in a balanced position to communicate within the earth as well as with spiritual realms. It has been recommended it be used to align the skeletal system, especially the spine. It assists in mending bone breaks and fractures and is used by healers as a treatment for all of the muscles and bones in the body.

ASTROLOGICAL: *Aries,* **CHAKRA:** *Solar Plexus,* **ELEMENT:** *Earth*

AZURITE

FAMILY: Carbonates

CHEMISTRY: $Cu_3(CO_3)_2(OH)_2$

HARDNESS: 3.5–4

STREAK: Light Blue

SG: 3.82–3.84

STRUCTURE: Monoclinic Crystal

AZURITE CRUST

AZURITE is one of the two basic carbonate minerals formed from the oxidation of copper; the other being malachite. The two are closely linked by chemistry and habit, and in fact, azurite will eventually become malachite upon further oxidation. Miners rely on the fact that azurite *(and malachite)* deposits are a good indication of the presence of copper-ore. In ancient times, azurite found in France, was traded throughout Europe and was highly valued for ornamentation. The Greeks and Romans used pulverized forms of the stone for medicinal purposes and as a coloring dye. In Asia and the Middle-East, lapis lazuli *(Lazurite)* was the preferred source for blue dye, since it was more readily available from abundant sources in Afghanistan. Like malachite, azurite is soft and damages easily. It does not occur in the massive, botryoidal or stalagmitic crystal forms like malachite does, so is not found as a tumbled and polished stone. Raw specimens are typically crusts on copper bearing rocks and are rarer and more costly than malachite or most other carbonates. Azurite is one of the purest blue-colored minerals and it makes a beautiful, lively and treasured addition to any collection.

Azurite releases the user from stress and worry, grief and sadness. It helps open the spirit, bringing light into the consciousness, bringing joy and positive energy to daily life. It helps with the release of fear and phobias, and helps change patterns of negative behavior that come from those insecurities. For relief of arthritis and joint pain, healers suggest placing azurite on affected areas of the body. It is also a useful stone for cleansing and detoxifying the body's systems.

ASTROLOGICAL: *Sagittarius*, **CHAKRA:** *Third Eye*, **ELEMENT:** *Water*

Related entries in this book: , Chrysocolla, Malachite

BARITE

FAMILY: Sulfates

CHEMISTRY: $BaSO_4$

HARDNESS: 3

STREAK: White

SG: 4.20–4.22

STRUCTURE: Orthorhombic Crystal

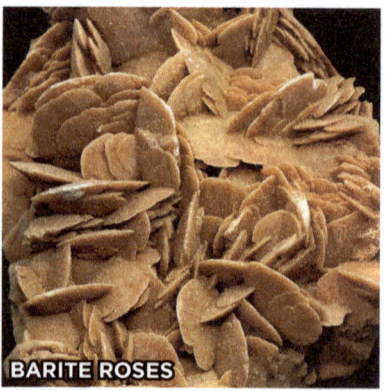
BARITE ROSES

BARITE is one of the most common sulfates on Earth. It occurs commonly in vein deposits, in association with several metals including hematite, lead and zinc. It can be found in limestones, and is frequently a binding agent in sandstones. Barite can form in hydrothermal zones and accumulates in hot spring deposits. It is very dense and heavy for a non-metal. Barite is also one of a few minerals that form attractive rosette-shaped crystal masses when subjected to repeated saturation and drying from groundwater. It has many commercial uses, such as a barrier-mud for oil drilling, as a filler for papermaking and as a shield against gamma radiation from equipment in scientific and hospital labs. China and the U.S.A. are the major producers of commercial barite. Barite is a very soft mineral that cracks and flakes easily, so it is not usually found as a tumbled and polished stone. Most specimens available to collectors are the attractive orange and white bladed-crystal clusters, or as "rosettes". Specimens as pictured above can be sought-out from better mineral and rock retailers or distributors. For such a common mineral, barite clusters make quite an unusual and attractive display piece in any collection.

Barite is a great stone to use when focusing energies on healing the Earth. It opens one's eyes to possibilities within us that are not always obvious. It promotes companionship, harmony, and love. It is of use to relieve all types of physical aches and pains and it helps detoxify the body making it especially useful for those trying to break addiction cycles.

ASTROLOGICAL: *Sagittarius,* **CHAKRA:** *Third Eye,* **ELEMENT:** *Earth*

BARITE BLADES ON GALENA

BERYLS

The family of beryls is one of the most widely-known and widely-used families of gem crystals. Some are commonly known, such as emeralds and aquamarines. Others less known, like goshenite and morganite. All are composed of the same basic silicate of beryllium and aluminum molecular structure. Minor impurities of other minerals are responsible for the varied array of hues found in beryls. They are hard, colorful, valuable, and extremely beautiful. Because they are found in so many locations worldwide, they have been a part of human history for millennia.

SEE PAGE #S BELOW FOR INDIVIDUAL BERYLS

BISMUTH

FAMILY: Native Element (Atomic #83)

CHEMISTRY: (Bi)

HARDNESS: 2–2.5

STREAK: Silver-White

SG: 9.68–9.74

STRUCTURE: Monoclinic Crystal

BISMUTH is considered to be the heaviest *(highest atomic mass)* stable, natural element, yet was recently discovered to be very slightly radioactive. However, it's only isotope, bismuth-209 decays with a half life more than a billion times the estimated age of the universe so it is probably safe to still refer to it as "stable". Bismuth occurs as an ore with lead, tin and sometimes silver deposits. It is a very stable element whose main use is in the production of cosmetics and as an antacid ingredient. Because it has many properties similar to lead, it can be used as a safe substitute for that metal in many industrial applications. Bismuth crystals, as pictured here from our family collection, do not occur in nature. They are made for the collector's market by a melting/cooling, precipitation cycle, mainly from a few reputable and high-quality sources in Germany. It has wonderfully complex and attractive stair-step crystals with sharp right-angles and is highly iridescent, displaying a full spectrum of rainbow hues. Small clusters of these crystals are easy to find on the market and are fairly brittle, but very attractive. The complex structure and shiny, rainbow coloring makes bismuth crystals a favorite with children as well as adults. In fact, it is usually one of the first specimens visitors will approach in our mineral specimen cabinets.

Bismuth relieves symptoms of isolation and loneliness, both spiritually and emotionally. It calms and directs personal change in a positive direction. It can transfer energy from the crown chakra to the base chakra, which is a unique trait,and it helps with attaining vision and wisdom. Physically, healers suggest bismuth for curing gastrointestinal disorders and disease *(External Use Only!)*.

ASTROLOGICAL: *None,* **CHAKRA:** *All,* **ELEMENT:** *Air*

BOJI STONE™

FAMILY: Mixed Mineral Concretion (mostly pyrite)

CHEMISTRY: FeS_2

HARDNESS: 6–6.5

STREAK: Brown/Black

SG: Varies

STRUCTURE: Mixed Crystal Structure

BOJI STONE™

BOJI STONE™ is a registered trademark for a concretion stone brought to market from a specific seller in Kansas. Similar concretions may occur nearby in a large, layered chalk formation, and are usually found in the layers of bentonite clay between chalk layers. These concretions, including the Boji stone, are also commonly known as Kansas pop-rocks. These hard, metallic-brown stones consist mainly of pyrite, marcasite, or other iron sulfates. Because of the inconsistent structure that is typical of most concretions, there are frequently small gaps in the internal structure of the stone. This, it was discovered, makes them explode if heated or thrown into a fire, earning them the name "pop-rocks". Some Boji stones display well-defined polycuboidal pyrite crystals on their surface and are referred to as "male" stones *(see photo)*. Smoother Boji stones are referred to as female. Because of their limited supply, Boji stones are fairly rare and are priced accordingly, however fine examples of other Kansas pop-rocks are sometimes available from reputable dealers at more affordable prices. It is common for them to be sold as a pair of male and female stones. Whether Boji stone or other pop-rock, they are a fascinating little item for the collection, having both visual appeal and unusually heavy weight.

Boji Stones are powerful and highly energetic. They greatly enhance spiritual energy, making them useful for any meditation or metaphysical endeavor. They bring clarity of thought and mental acuity, focusing inner-vision and intuition. Physically, Boji stones have been recommended for improving balance and coordination. They are considered to be an aid for almost any type of physical healing.

ASTROLOGICAL: *Scorpio*, **CHAKRA:** *All*, **ELEMENT:** *Earth*

Related entries in this book: Moqui Marbles, Prophecy Stone, Septarian,

BORNITE (Peacock Ore)

FAMILY: Sulfides

CHEMISTRY: Cu_5FeS_4

HARDNESS: 3-3.25

STREAK: Grayish Black

SG: SG 5.08-5.11

STRUCTURE: Orthorhombic Crystal

BORNITE

BORNITE (PEACOCK ORE) and chalcopyrite are both commonly referred to as peacock ore. Both are sulfides of copper and are commonly found in association with high-grade copper ore deposits. Both varieties may display the typical, multi-colored luster, however in the commercial market, chalcopyrite is most often acid treated to oxidize the copper content and bring out its colors. The two can usually be distinguished by chipping or scratching through the surface. Chalcopyrite should appear metallic and golden-yellow below the surface, while bornite will have pinkish or colorless stony texture. The colors of natural chalcopyrite also fade to black rather quickly, which is why it is so often acid-treated. Since this is a very common substance to begin with, acid-treating is not considered to decrease the value of the stone and it will ensure that its beauty remains intact. Peacock ore has long been a favorite with collectors as it is accessible, inexpensive, and a bright and colorful addition to any collection. A reputable and knowledgeable dealer should be able to lead you to a natural and non-treated specimen of bornite peacock ore for your collection.

ASTROLOGICAL: *Sagittarius,* **CHAKRA:** *Solar Plexus,* **ELEMENT:** *Fire*

Related entry in this book: Chalcopyrite

BRONZITE

FAMILY: Inosilicates

CHEMISTRY: $(Mg,Fe)SiO_3$

HARDNESS: 5.5

STREAK: White

SG: 3.20–3.21

STRUCTURE: Orthorhombic Crystal

BRONZITE

BRONZITE is an orthopyroxine mineral found in both igneous or metamorphic rocks. It is a member of the enstatite and ferrosilite series. At one end, enstatite is high magnesium/low iron, and at the other end of the series, ferrosilite is high iron/low magnesium. Hypersthene has a significant amount of both elements. Bronzite is most similar to hypersthene, but has a significantly higher ratio of iron. This is responsible for its characteristic dull brownish-bronze hues, with its black inclusions being specks of hematite or goethite. While not nearly as popular or attractive as hypersthene, and lacking the shiller *(shimmer effect)* of hypersthene, bronzite is still used as a decorative and ornamentation stone. It can be found in several locations throughout Asia and Europe, and beautiful, translucent, brown crystals have been found in New York, Pennsylvania, Texas and Oregon in the United States. Bronzite crystals are rare, but nice tumbled stones can be found with a little hunting at mineral and rock shops.

Bronzite offers protection to the bearer, as it repels and sends back negative energies to the sender. Spiritually, bronzite assists in achieving a state of certainty without doubt, allowing one to easily adapt to the best path to take. It increases assimilation and retention of iron in the body. It aids the body in the normalizing body cycles and regulates the internal clock.

ASTROLOGICAL: *Leo,* **CHAKRA:** *Base,* **ELEMENT:** *Earth*

Related entry in this book: Hypersthene

BUMBLEBEE

FAMILY: Mixed Mineral

CHEMISTRY: $CaCO_3 + As_2S_3$

BUMBLEBEE

BUMBLEBEE is commonly referred to as "Bumblebee Jasper", however there is nothing Jasper about it. It is not a silicate at all but a mixed-mineral rock composed primarily of bands of multi-colored calcite, with bright yellow bands pigmented with realgar *(arsenic)*. Because of the arsenic content and the softness of the rock, care should be taken handling it. Please wash hands after handling! However, bumblebee can be professionally stabilized by various making it bitch safe to handle and suitable for cutting for jewelry purposes. Currently, bumblebee is only sourced from only from Indonesia, as no other deposits are known at this time.

For healing purposes, bumblebee has been suggested for energy, vibrancy, and increasing personal activity levels. It is a beautiful stone that will undoubtedly make its way into the hands of many crystal healers.

ASTROLOGICAL: *Leo,* **CHAKRA:** *Sacral,* **ELEMENT:** *Fire*

Related entries in this book: Calcite, Orpiment, Realgar

BUTTERSTONE

FAMILY: Silicates

CHEMISTRY: SiO_2
(Silicified Organic Matter)

HARDNESS: 7.0

STREAK: White

SG: 2.58–2.59

STRUCTURE: Amorphous
Crystal Structure

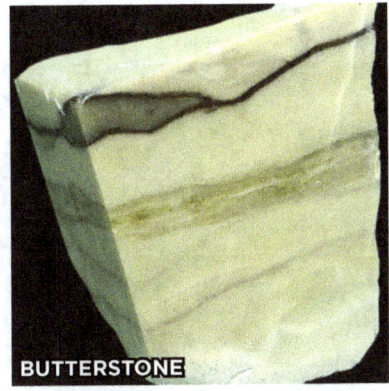
BUTTERSTONE

BUTTERSTONE is properly known as archaen butterstone, and is an organic-derived rock containing the silicified* remains of ancient blue-green algae. It is predominantly found in ancient, continental greenstone belts.** In regions of South Africa, these algae deposits date back approx. 2-1/2 billion years to the archaeozoic era, making them some of the very earliest forms of life on earth. By virtue of its cryptocrystalline silica structure and the fact that its defining coloring and pattern came from organic matter, it is similar in that respect to calligraphy stone, turritella agate, petrified wood and several jasper varieties.

Butterstone is used as a stabilizing healing stone, providing grounding and direction. In stressful times it can bring back focus, determination and can instill hope in times of despair. Because of its smooth "buttery" texture, it is most beneficial to hold this stone when using it as a healing Crystal.

ASTROLOGICAL: *Sagittarius,* **CHAKRA:** *Heart,* **ELEMENT:** *Earth*

(silicification)—Some minerals or organic materials undergo a process called pseudomorphism, which means that their original chemical structure has been replaced by another mineral. When matter is replaced by quartz or other silicate it is said to have "silicified".

** (greenstone belt)—Derives its name from the color imparted to the mineral bed by its chloride content. These rock formations are typically mafic (magnesium and iron) in composition, indicating an origin as a sea floor or from deep in the Earth's crust. Compressed over time, and moved landward by tectonic plate movement, greenstone belts are usually indicators of the oldest parts of a continent.*

CALCITE

BROWN DOGTOOTH CALCITE

GREEN CALCITE

CALCITE

FAMILY: Inosilicates

CHEMISTRY: $CaCO_3$

HARDNESS: 5.5

STREAK: White

SG: 3.20–3.21

STRUCTURE: Orthorhombic Crystal

BLUE CALCITE

CALCITE is a both mineral and a name of a mineral group. The calcite group are minerals containing a carbonate compound plus a metal, and include: magnesite *(magnesium)*, siderite *(iron)*, rhodocrosite *(manganese)* and dolomite *(calcium magnesium)*. Calcite itself is one of the most common minerals in the world and forms rocks such as limestone, marble, chalk and travertine. Calcite occurs in a huge range of colors, and it is found almost everywhere on Earth in some form. It can occur in veins, fissures, hydrothermal zones, marine environments, inside geodes and even as mineral deposits on faucets, sinks and tubs in the house. It is soft but forms attractive crystals, and because of its abundance and variety, calcite is found in most rock collections and is quite affordable.

BLUE CALCITE is an excellent stone for students. It helps retain lessons learned and amplifies new knowledge. It is soothing and calming. Blue calcite will stimulate the metabolism, strengthen the immune system, stabilize the rhythm of the heart, which will also lower blood pressure. It is useful for alleviating pain.

ASTROLOGICAL: *Virgo,* **CHAKRA:** *Base,* **ELEMENT:** *Earth*

BROWN DOGTOOTH CALCITE is one of the more unusual and attractive crystalline forms of calcite and is excellent for display in a collection. It generally forms inside geodes and cavities, and can display a wide range of colors. Dogtooth calcite helps retain memory of astral-traveling and energy-channelling experiences. It enhances and enables deep meditation if held or placed nearby. It balances the metabolism, reduces nervous energy and calms hyperactivity in children.

ASTROLOGICAL: *Cancer,* **CHAKRA:** *Third Eye,* **ELEMENT:** *Fire*

CALCITE FAIRY STONE

GREEN CALCITE

OPTICAL CALCITE
(Iceland Spar)

GREEN CALCITE balances and attunes the body's electrical circuitry, bringing better harmony between mind and body, allowing one to reach a higher potential of focus and coordination

ASTROLOGICAL: *Cancer,* **CHAKRA:** *Throat,* **ELEMENT:** *Earth*

CALCITE FAIRY STONE

These formations of calcite display a wide range of shapes that can look like they were carved, or molded making them popular with collectors. For crystal healing, calcite fairy stones are commonly carried as an amulet to ward-off negative energy, but they also promote open-mindedness and clarity of thought. They make a nice addition to any collection.

ASTROLOGICAL: *Taurus,* **CHAKRA:** *Base,* **ELEMENT:** *Earth*

OPTICAL CALCITE *(Iceland Spar)* occurs in large crystals and cleaves easily into rhombus shapes. It is also remarkable for its double refraction of light *(bi-refringence)*. Optical calcite clears away negative energy, and amplifies positive energy. Specifically, this mineral helps remove fear-based emotions, bringing more optimism into one's experiences. Great for bones and joints, and balancing the amount of calcium in the body. It also helps the body to absorb important vitamins and minerals.

ASTROLOGICAL: *Leo,* **CHAKRA:** *Crown,* **ELEMENT:** *Earth*

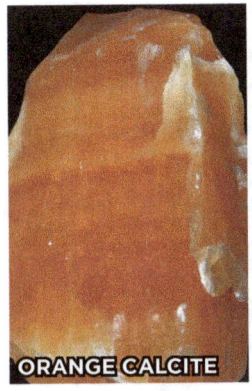

ORANGE CALCITE   RED CALCITE   YELLOW CALCITE

ORANGE CALCITE boosts energy, cleanses the aura, brings prosperity, reduces fear and reduces stress. A great all-purpose healing crystal! Good for back pain and increasing physical strength. It is used to treat tooth problems and optical disorders. A good detoxifying and antiseptic agent, orange calcite helps to heal intestines and irritable bowel syndrome, kidneys, and chronic fatigue. It helps increase calcium intake and aids with assimilation of minerals into bones and teeth.

ASTROLOGICAL: *Aries,* **CHAKRA:** *Solar Plexus,* **ELEMENT:** *Earth*

RED CALCITE helps dispose of negative energy and emotions from the body and mind, and releases fear. It can be used for problem-solving and helps to find true love. Red calcite regulates adrenaline and promotes restful sleep and peaceful dreams. Physically it is useful for relieving aching hips, legs, feet and muscles.

ASTROLOGICAL: *Virgo,* **CHAKRA:** *Heart,* **ELEMENT:** *Earth*

YELLOW CALCITE stimulates the intellect and sharpens the memory. It can help one to organize intellectual thoughts and retain new information. Good for astral-projection and meditation. Physically, yellow calcite treats the stomach, intestines, upper back and the spine.

ASTROLOGICAL: *Aries,* **CHAKRA:** *Solar Plexus,* **ELEMENT:** *Earth*

CALLIGRAPHY STONE

FAMILY: Silicates

CHEMISTRY: SiO_3 + Fe_3

HARDNESS: 6.0–6.5

STREAK: White

SG: 2.65–2.90

STRUCTURE: Trigonal Crystal

CALLIGRAPHY STONE

CALLIGRAPHY STONE is also commonly called script stone, Miriam (or Miryam) stone and Arabic writing stone. It is a beautiful representation of ancient sea fossils embedded in dark hematite iron deposits, which together have been silicified* in the form of a hard chalcedony. It can be found in abundance in the Himalaya mountains where evidence of an ancient seabed lies high among the peaks of the mountain range. This striking inconformity is due to uplift of mountains, as the Indian sub-continent continues to subduct underneath the Asian continent. This collision began approximately 55 million years ago and continues to this day, raising the Himalayan mountain chain by 5 millimeters per year. The fossilized patterns of calligraphy stone are striking, unique and reminiscent of ancient writing styles, hence its name.

Calligraphy stone is a great aid for focused meditation and for those seeking wisdom and knowledge from the Akashic records. At the same time, this stone is of great protection to the spiritual self and helps ward-off unwanted energy influences.

ASTROLOGICAL: *Cancer,* **CHAKRA:** *Crown,* **ELEMENT:** *Earth*

(silicification)—Some minerals or organic materials undergo a process called pseudomorphism, which means that their original chemical structure has been replaced by another mineral. When matter is replaced by quartz or other silicate it is said to have "silicified".

CARBORUNDUM (Moissanite)

FAMILY: Artificial Mineral

CHEMISTRY: SiC

CARBORUNDUM

CARBORUNDUM (MOISSANITE) is composed of pure silica and pure carbon, and is a manufactured mineral not found in nature. It can be a by-product of various industrial processes but is also lab-grown for the collectors market. The mineral moissanite is identical in chemistry and can be found in nature, but does not display the color and striking structure of its manufactured relative. While it is very brittle and "flaky" when handled, it is actually extremely hard. Harder in fact, than diamond. *(Remember, "toughness" and "hardness" are not the same thing).* It's main industrial use is as an abrasive and can be used to polish other very hard substances like ruby and sapphire. Polishing disks and grinding tools commonly use carborundum surfaces.

Crystal healers suggest using carborundum for achieving goals and finding inner-strength to finish projects. Two friends or mates who each possess a carborundum piece are said to be inseparable for eternity!

ASTROLOGICAL: *Aquarius*, **CHAKRA:** *All*, **ELEMENT:** *Air*

CARNELIAN

FAMILY: Cryptocrystalline Silicates

CHEMISTRY: SiO_2

HARDNESS: 7

STREAK: White

SG: 2.58–2.59

STRUCTURE: Triclinic Crystal

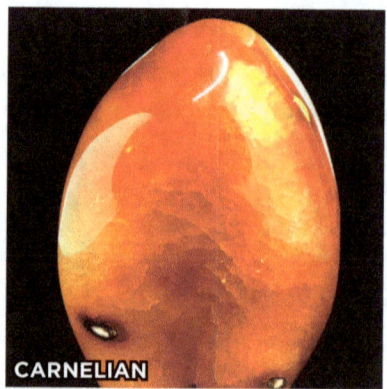
CARNELIAN

CARNELIAN is a beautiful and vibrant variety of chalcedony, usually gaining its color from impurities of iron oxide. While a true carnelian is usually semi-translucent and colored bright or deep orange, it can also be a mixture of orange with patches of colorless, yellow or brown chalcedony. Since the difference between forms of chalcedony is usually dependent on appearance *(banding, mottling, uniform)*, there are differing opinions as to whether carnelian is actually considered to be chalcedony, an agate or its own variety of cryptocrystalline quartz. Partly due to its historic uses and references, carnelian has retained its popular definition as a specific and unique mineral. Carnelian has been used as a semi-precious gemstone for jewelry and ornamentation for thousands of years.

Carnelian is a great stone for motivation and for boosting one's energy and spirits. It provides courage when making important life decisions, can change feelings of apathy and is a motivator for success. Great for musical and artistic inspiration, it can be kept nearby when giving public performances. Carnelian has uses as treatment for lower back problems, rheumatism, arthritis, neuralgia and depression. It regulates kidney function and accelerates the healing of bones and ligaments. Carnelian improves vitamin and mineral absorption and ensures good blood supply to organs and tissues.

ASTROLOGICAL: *Taurus,* **CHAKRA:** *Base,* **ELEMENT:** *Water*

Related entries in this book: Agate, Chalcedony, Jasper

CAVANSITE

FAMILY: Silicates

CHEMISTRY: $Ca(VO)Si_4O_{10} 4(H_2O)$

HARDNESS: 2.25–2.33

STREAK: Bluish-white to white

SG: 2.25–2.33

STRUCTURE: Orthorhombic Crystal

CAVANSITE ON STILBITE

CAVANSITE gets its name from its chemical components, calcium vanadium silicate. It is a deeply blue, silicate mineral which forms delicate radiating crystals and is rarely found in any sizable mass. It commonly occurs in cavities in basalt lavas such as in the Deccan Traps region in India, and is found in association with many other zeolite minerals. Recent expansion of mining efforts in this region have brought many new and interesting minerals to the general market, cavansite being one of the most unusual and beautiful. While the typical crystal displays are quite small, they are a fine example of a very unusual crystalline habit and are worthy of display space in any collection. A magnifier or lens is ideal for viewing the crystals. The close-up shot of the cavansite crystal shown here, is enlarged to approximately 3 times its actual size.

Cavansite is helpful when defining personal boundaries and for maintaining self-esteem. Past-life seekers use this crystal for clarifying past experience and relating to today's manifestation. Physically cavansite is used to prevent illness and maintain the body's defenses against disease. Cavansite is of help defining personal boundaries and maintaining self-esteem.

ASTROLOGICAL: *Aquarius,* **CHAKRA:** *Throat,* **ELEMENT:** *Water*

Related entries in this book: Scolecite, Stilbite

CELESTITE

FAMILY: Sulfates

CHEMISTRY: $SrSO_4$

HARDNESS: 3-3.5

STREAK: White

SG: 3.96-3.98

STRUCTURE: Orthorhombic Crystal

CELESTITE GEODE

CELESTITE also known as celestine, can usually be found in areas of lime-stone and sandstone deposits. It is also frequently found often in association with evaporate deposits. Crystals can grow in many forms from tabular, prismatic, nodular, or most commonly in hexagonal forms. Celestite is often found in geodes and has a beautiful transparent "celestial" blue tint due to the molecular content of strontium. Its unique translucent, blue coloring makes it extremely popular with collectors. Where it is abundant enough, celestine is mined for its strontium content. It is fairly common in small deposits worldwide. Madagascar is currently a major source of commercially available specimens.

Celestite is often suggested to elevate moods and dispel sadness. It helps to teach one how to avoid harmful behavior tendencies. It promotes sound and peaceful sleep, memorable dreams and mental clarity. Celestite eases stress in the body and is useful for preventing illness brought on by worry. It it used improve eyesight and hearing and relieves sore throats by placing the crystal on the afflicted area *(External Use Only!)*. Celestite helps functions of the thyroid gland, relieves digestive problems and helps eliminate toxins from the body. In the crystal healing realm, celestite is often referred to as angelite, however mineralogically that name is reserved for the light blue variety of anhydrite. See the listing in this book for anhydrite for more information about this non-related mineral.

ASTROLOGICAL: *Gemini*, **CHAKRA:** *Throat*, **ELEMENT:** *Earth*

CHALCEDONY

FAMILY: Cryptocrystalline Silicates

CHEMISTRY: SiO_2

HARDNESS: 7

STREAK: White

SG: 2.58–2.59

STRUCTURE: Trigonal Crystal

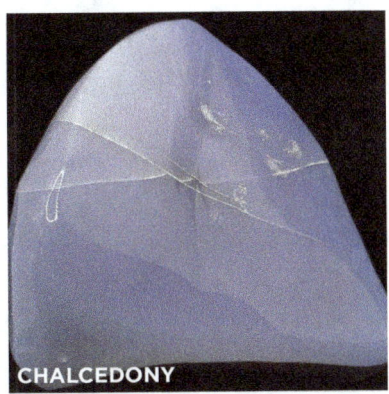
CHALCEDONY

CHALCEDONY is the cryptocrystaline *(microscopic)* silicate that defines a wide family of quartz minerals; agate, jasper, carnelian, onyx, heliotrope, aventurine and chrysoprase. Translucence, banding, habit, and occurrence, are all factors that determine the specific variety. Chalcedony is typically light blue or gray, occasionally yellow, and is often the mineral crust that holds geodes together. Because of its beauty, abundance and durability, chalcedony has been used for household and decorative purposes for many millennia.

Chalcedony is a stone of cooperation and balance between the self and others. It promotes brotherhood, peace and good will. It absorbs negative energy and deflects negative influence, bringing the mind, body and spirit into harmony without distraction. It can help promote better absorption and incorporation of minerals ingested. Chalcedony keeps the mind sharp and decreases the effects of dementia and senility. It increases and sustains physical energy. It heals ailments of the eyes, gallbladder, bones and spleen, and fortifies the blood and the circulatory system.

ASTROLOGICAL: *Cancer,* **CHAKRA:** *Sacral,* **ELEMENT:** *Earth*

Related entries in this book: Agate, Jasper, Carnelian, Onyx, Heliotrope, Aventurine, Chrysoprase

BLUE CHALCEDONY

YELLOW CHALCEDONY

CHALCOPYRITE

FAMILY: Sulfides

CHEMISTRY: Cu_5FeS_4

HARDNESS: 3–3.25

STREAK: Grayish Black

SG: SG 5.08–5.11

STRUCTURE: Orthorhombic Crystal

CHALCOPYRITE ON MATRIX

CHALCOPYRITE is a very commonly collected mineral and has also become a popular commodity for healers. Like bornite, chalcopyrite is often treated to enhance color and reduce fading. Bright, rainbow-colored specimens referred to as "peacock ore" are abundant and extremely popular with collectors of all ages. Surface oxidation of the copper content in chalcopyrite is the cause for the coloring, so any freshly broken or fractured samples will not show color on the inside surfaces. Occasionally, larger crystal forms of chalcopyrite can be found at mineral dealers, but rarely at healing centers or gift shops. They display wonderful, metallic and semi-cubic crystals that are appreciated by mineral collectors for their form, regardless of their bland appearance *(photo above)*.

Chalcopyrite can be used to balance adrenaline levels in the body and stabilize bodily energy. It improves cell regeneration, and is used to help reduce fevers and inflammation. Peacock ore grounds excess nervous energy, allowing the body to release stress and calms emotions which cause imbalance throughout the body.

ASTROLOGICAL: *Capricorn,* **CHAKRA:** *Base,* **ELEMENT:** *Fire*

Related entry in this book: Bornite

CHAROITE

FAMILY: Silicates

CHEMISTRY: $K_5Ca_8(Si_6O_{15})_2$

HARDNESS: 5-6

STREAK: White

SG: 2.54–2.58

STRUCTURE: Monoclinic Crystal

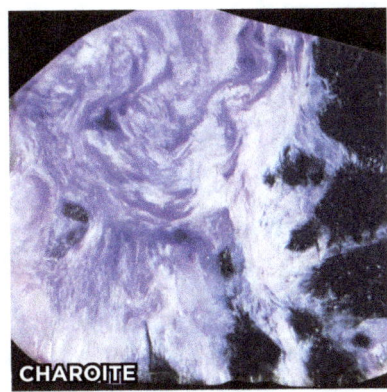
CHAROITE

CHAROITE is a fairly rare silicate mineral, and while it was apparently discovered in the 1940's it has only been widely, commercially available for the past two decades. It has been found in only one location in Siberia, Russia and is named for the Chara River which flows nearby its source. This source lays within in an intrusion of an igneous rock called syenite; similar to granite but with lower quartz content. It has an unusual and pleasant, swirling, fibrous appearance, and is sometimes slightly chatoyant *(cat's-eye effect)*. That, along with its intense purple color *(from its potassium content)* and a pearly, almost "plastic" appearance, can lead many to believe it is synthetic or enhanced artificially. The first samples of charoite that were found seem to be of a higher quality and richer color than more recently mined pieces, however fine specimens are still circulating through the commercial market. Charoite has become very popular for use in jewelry, and its price has increased significantly. Charoite would be a unique and special addition to any rock collection.

Charoite is a powerful stone for transformation. It cleanses the aura and chakras by changing negative energy into healing energy. It has the strong power to give a sense of purpose to daily functions, and give the ability and foresight to prioritize tasks that should be addressed seriously. Physically, charoite stimulates and regulates blood pressure which improves sleep habits and makes a well-rested mind. Charoite specimens on the market today are usually relatively small but it has a good reputation as being a powerful healing stone regardless of size. A good clear quartz crystal can greatly amplify the effectiveness of charoite.

ASTROLOGICAL: *Scorpio*, **CHAKRA:** *Heart*, **ELEMENT:** *Air*

CHRYSOBERYL & CHRYSOBERYL CAT'S EYE

FAMILY: Oxides

CHEMISTRY: $BeAl_2O_4$

HARDNESS: 8.5

STREAK: White

SG: 3.75

STRUCTURE: Orthorhombic

CHRYSOBERYL

CHRYSOBERYL an extremely hard precious gemstone and has been used in jewelry for many centuries. One variety, "alexandrite" was found in Russia and named for Tsar Alexander in celebration of his birthday. Prized Alexandrite specimens are highly pleochroic and display color-change from red to green. Common varieties of chrysoberyl range from yellow to green to gold. Cymophane *("Cat's Eye")* is a unique variety of chrysoberyl that displays chatoyance *(similar to asterism, or the play of color found in tiger's eye stones)*. As a gemstone chrysoberyl is not as commonly sold as emerald, ruby or sapphire but is quite beautiful and takes faceting extremely well.

For healing crystal purposes, Chrysoberyl of all varieties are suggested for focus, concentration and spirituality. Cat's Eye specifically, is used to ward off evil or negative energy.

ASTROLOGICAL: *Leo,* **CHAKRA:** *Heart,* **ELEMENT:** *Water*

CHRYSOBERYL CAT'S EYE

CHRYSOCOLLA

FAMILY: Oxides

CHEMISTRY: $(Cu,Al)_2H_2Si_2O_5(OH)_4nH_2O$

HARDNESS: 2.5–3.5

STREAK: White to Blue-Green

SG: 2.00–2.40

STRUCTURE: Orthorhombic Crystal

CHRYSOCOLLA

CHRYSOCOLLA is a Hydrated copper cyclosilicate mineral that commonly forms in copper ores. It can form as botryoidal lumps, rocky masses or small rock clusters. It most often located in association with azurite and malachite, due to their similar chemistry and is often hard to distinguish among them. Chrysocolla is fairly soft and chips very easily. Examples of chrysocolla that can be used for polishing, carving or for jewelry are generally an agatized or opalized form of chrysocolla, meaning that some quartz has either intergrown with the oxidized copper crystals, or the chrysocolla has been entirely or partially pseudomorphed* into quartz. Chrysocolla is a very colorful stone that has been popular throughout history, wherever it is found.

Chrysocolla improves communication and promotes openness to psychic vision, self-awareness and inner balance. It brings confidence, yet helps one to maintain sensitivity. It enhances personal power and is a strong inspiration for creativity. Physically, healers suggest using chrysocolla as an aid for arthritis, bone disease, muscle spasms, blood disorders and lung problems. It is also used to treat the liver, kidneys and intestines, and to treat infections. It can be used for high blood pressure and as an aid to heal burns.

ASTROLOGICAL: *Taurus,* **CHAKRA:** *Throat,* **ELEMENT:** *Earth & Air*

Related entries in this book: Azurite, Malachite

*(pseudomorphism)—The process of one mineral replacing another mineral chemically, yet leaving the crystal structure, appearance, and usually the color intact.

CHRYSOPRASE

FAMILY: Cryptocrystalline Silicates

CHEMISTRY: SiO_2

HARDNESS: 7

STREAK: White

SG: 2.80-2.84

STRUCTURE: Monoclinic Crystal

GREEN CHRYSOPRASE

CITRON CHRYSOPRASE

CHRYSOPRASE is a variety of chalcedony that contains small quantities of nickel, which typically give it a light, minty-green coloration. It commonly occurs as an end-product of nickel-rich silicate being exposed to weathering. It is often found mixed with small veins of iron oxides which are seen as rusty, brown streaks. Uniformly rich green specimens are valued as gemstones for jewelry use. Chrysoprase also occurs in a creamy, "citron" color that is far less common, but perhaps even more treasured for ornament. Citron chrysoprase differs in its entire structure and appearance because it started as magnesite that pseudomorphed* into chrysoprase by silicate replacement. Currently, Russia, Germany and the United States are the major sources of chrysoprase. Australia remains the prime source for citron or lemon varieties. Non-gem quality specimens of all types of chrysoprase are fairly common and inexpensive and can be easily found in polished or rough form.

Chrysoprase has tranquilizing effects on a busy mind. It helps to deal with issues which demand clear-headed decision making. It calms irritability and protects the bearer from bad dreams. Healers suggest chrysoprase for improving eyesight and relieving the pains of rheumatism and arthritis. It is also useful for healing any type of external wound, and is effective against all types of allergy symptoms.

ASTROLOGICAL: *Cancer,* **CHAKRA:** *Heart,* **ELEMENT:** *Earth*

Related entry in this book: Chalcedony

(pseudomorphism)—The process of one mineral replacing another mineral chemically, yet leaving the crystal structure, appearance, and usually the color intact.

CINNABAR

FAMILY: Sulfide

CHEMISTRY: HgS

HARDNESS: 2.0–2.5

STREAK: Rhombohedral

SG: 8.2

STRUCTURE: Trigonal

CINNABAR

CINNABAR is the crystalline form of mercury and as such, raw specimens should be handled with great care, or not at all. If contact is made with raw crystals, hands should be washed immediately. Excellent specimens of cinnabar crystals are commercially available and are actually quite popular due to their vivid red color. Small and formless masses are most common, however decent crystal prism shapes can be found. More commonly, small, red crystals of cinnabar in a matrix of white quartz are sold as tumbled stones. These can be handled safely. Tumble-polished stones of cinnabar in quartz sold commercially can provide an inexpensive and safe way to add mercury crystals to your collection.

There is evidence of cinnabar being mined in hydrothermal locations for at least 11 Centuries! The deep red pigment "vermillion" comes from the cinnabar mineral and was quite commonly applied in ancient Chinese artworks and statuary, as well as later Greek and Roman use.

As for crystal healing, please heed the warning again, that raw cinnabar crystals are extremely toxic to the nervous system and should never be handled. Tumble-polished stones are more appropriate. Cinnabar has long-been used as a crystal for manifesting love or enhancing personal attraction. Some use it to attract wealth and power. In my research I have not found any reliable source that indicates cinnabar has ever been used for physical healing.

ASTROLOGICAL: *Cancer,* **CHAKRA:** *Base,* **ELEMENT:** *Fire*

CITRINE

FAMILY: Quartz Silicate

CHEMISTRY: SiO_2 (with Traces of Fe)

HARDNESS: 7

STREAK: White

SG: 2.65–2.68

STRUCTURE: Trigonal Hexagon Crystal

CITRINE

CITRINE, like its closely related cousin amethyst, it is a popular semi-precious gemstone and has been used for ornamentation for thousands of years. Like purple amethyst, citrine gains its rich color from traces of iron ions and exposure to natural irradiation. Citrine forms in a wide range of generally yellow or citrus colored hues. We have specimens in our family collection that are bright and richly golden, and others that are almost amber or honey colored. As is true with all types of quartz, it is very hard and transparent and facets well, making it ideal for jewelry settings. It is widely available commercially, and some of the finest gem-quality crystals rival much more expensive jewels for beauty and luster. Small clusters and single, crystal points are usually offered on the retail market for collectors and healers.

Citrine energizes all levels of being. It cleanses the chakras and expands intuition. It represents wealth and success and helps with focus and concentration. Physically, healers use citrine to treat degenerative diseases. It is used for improving digestion and the health of the spleen and pancreas. Citrine treats kidney and bladder infections and constipation, and is also a treatment for cellulite.

ASTROLOGICAL: *Aries,* **CHAKRA:** *Sacral,* **ELEMENT:** *Fire*

Related entries in this book: Amethyst, Quartz

CITRINE

CONCRETIONS

A concretion is an assemblage of discrete minerals, formed into one solid formation, or "rock". They can be multi-layered with one or more mineral surrounding others in a sphere or orb *(e.g. Moqui Marble)*, or can be fractured solids that later filled-in with other minerals *(e.g. Septarians)*. Concretions can be found in-situ, embedded in host rocks *(usually sedimentary)*, or floose, after weathering out of a solid rock bcd or geologic formation.

Concretions should not be confused with geodes which can have a similar outward appearance and can form in similar geologic conditions. Geodes may also be composed of several discrete minerals, but they form crystal-linings over long periods of time, in liquid-filled vugs *(voids)* in their host rocks, with different chemistry of the liquids determining the type of crystals that form inside the geode.

SEE PAGE #S BELOW FOR INDIVIDUAL CONCRETIONS

COPAL

FAMILY: Polymerized Tree Resin

HARDNESS: 1-1.5

STREAK: Same Color as Specimen

SG: 2.80-2.84

STRUCTURE: Organic Rock

COPAL

COPAL is tree resin that has polymerized but not yet had enough time to fossilize completely, like amber. It can have properties and appearance similar to amber, even inclusions of insect and plant life. It is commonly found in Mexico, Columbia and several locations in Central America. Softer formations of Copal have been burned as incense for ceremonial purposes for hundreds, perhaps thousands of years by meso-Americans. Harder amber is more desirable for cutting into jewelry grade decoration. East Africa was once the source of much of the world's copal and it was used extensively in Britain and Europe as a source of natural varnish. Copal can usually be distinguished from amber by its lighter color, its softness and its reaction to solvents. Mineralogically, copal is categorized as an organic rock, not a mineral.

Copal is a powerful aid for self-healing, cleansing the spirit, and clearing the mind. It brings good luck and good fortune. Copal is used by healers to treat ailments of the internal body systems and most specifically the liver, kidneys and bladder. It is also useful as treatment for sore throats and laryngitis.

ASTROLOGICAL: *Leo,* **CHAKRA:** *Solar Plexus,* **ELEMENT:** *Earth*

Related entries in this book: Amber, Jet

COPPER

FAMILY: Native Element (Atomic #29)

CHEMISTRY: Cu

HARDNESS: 2.5–3

STREAK: Rose Pink

SG: 8.96–8.97

STRUCTURE: Isometric Crystal

FLOAT COPPER

COPPER is such an important and abundant element and is often overlooked by mineral collectors. Native copper takes many varied shapes that make it attractive for display. Float copper and splash copper *(see photos)* are particularly attractive and are formed by pouring molten copper into vats of water, partially containing layers of straw or other material. The hot copper cools an hardens quickly in the water. Float copper flattens and hardens instantly, close to the surface of the water, spreading-out into beautiful free form plates. Splash copper drips below the surface and makes dendritic displays. At least some form of copper can be found at almost all collector or crystal-healing outlets, and consideration should be given to obtaining some pieces of this fine metal.

Copper is a conductor of spiritual and physical energy, and is useful for a multitude of purposes to the healer or crystal user. Copper is used to balance the chakras and meridians and for improving coordination, balance and physical strength. Copper aids in transmitting experience from other realms to the conscious and subconscious mind. It brings luck and good fortune to the bearer. Copper is of great use to those suffering from arthritis, bursitis or maladies of the joints and is most effective when worn or placed directly on the affected areas.

ASTROLOGICAL: *Taurus,* **CHAKRA:** *Base,* **ELEMENT:** *Fire*

SPLASH COPPER

CORAL

FAMILY: Organic Material

CHEMISTRY: Mostly $CaCO_3$

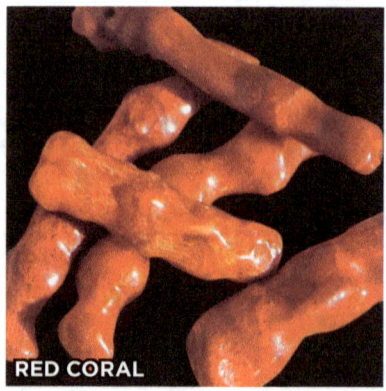

RED CORAL

CORAL is not a mineral, but an organic form of sea-life. It is so abundant and found in so many mineral and rock collections, that we decided to include it in this book. Whether recent or fossilized, corals have been used around the world as decoration, ornament and tools for thousands of years. Its many colors and great availability, make it easy and inexpensive to collect. Red coral jewelry beads have become a very popular decoration the past few decades. While some examples are clearly dyed to unrealistic shades of red, true coral has its own rich and bright hue.

Red Coral gemstone is used for enhancing self-confidence, organization and to negate the power of bullying or detrimental actions. This makes it a very desirable protection stone and is easily worn as a necklace, pendant, ring or bracelet. Healers describe coral as a potent aid against fever, impotency, and stomach aches.

ASTROLOGICAL: *Aries,* **CHAKRA:** *Crown,* **ELEMENT:** *Water*

WHITE CORAL

CORUNDUM

FAMILY: Oxides

CHEMISTRY: Al_2O_3

HARDNESS: 9 (defining)

STREAK: White

SG: 3.99–4.01

STRUCTURE: Hexagonal Crystal

RUBY

CORUNDUM minerals are ruby and sapphires, which are identical aluminum oxides. Rubies are shaded red from chromium content. Sapphires get their blue color from aluminum and titanium content in the stone. All other colors of corundum are referred to as sapphires, and take their specific names from their coloring, which comes from trace-content of other metals. All forms of corundum are found in only a very few places in the world including Burma, Myanmar, Columbia, Sri Lanka, Madagascar, Australia and Kenya. Sapphires and rubies like many precious gemstones, are formed in pegmatite intrusions of igneous rock such as granite or rhyolite. However, most sapphires and rubies are collected from alluvial deposits at, or just below ground level, both on dry land and in river-beds. Alluvial deposits are formed by collections of sediments and minerals that have been carried downhill by water-erosion of mountains or hills. The process for collecting the gemstones from these deposits is very similar to panning for gold. Of all the sapphires and rubies found, only a very small fraction are fine enough quality make it to the gem market. Most gem quality stones are heat treated to bring out the richness of color. With very few, rare exceptions, untreated sapphires and rubies can be dull and lifeless with very little brilliance, so heat-treating has long been an accepted industry practice.

BLUE SAPPHIRE

CREEDITE

FAMILY: Halides

CHEMISTRY: $Ca_3Al_2SO_4(F,OH)_{10}+2(H_2O)$

HARDNESS: 4

STREAK: White

SG: 2.69–2.71

STRUCTURE: Monoclinic Crystal

CREEDITE

CREEDITE is a fairly uncommon hydroxyl halide mineral. It usually forms from the oxidation of fluorite in ore deposits. It was named after the location where it was discovered in 1916 at a fluorite mine near Creede Quadrangle, Colorado. Its clusters of sharp-bladed and needle-like crystals, and bright orange color make it instantly recognizable. It is brittle and fairly soft, and its crystal-line composition make it unsuitable for tumbling, polishing, faceting or cutting. Recently, a limited supply of fair-sized and beautiful creedite clusters have made their way to the collector's market from Mexico. We have two fine specimens in our family collection, and the combination of the striking crystal display and bright coloring make it a favorite for viewing.

Creedite is used when trying to clarify life-lessons and assimilate their meaning, including deciphering experiences from past lives. It allows one to view and judge circumstances without bias, and to remain centered and focused even in stressful or difficult circumstances. Creedite is a powerful tool when used in a crystal grid or placed centrally in the home to offer safety and protection to the household. Physical uses for creedite include strengthening the immune system, and helping the body to absorb nutrients and vitamins. It can help to speed recovery from fractures, torn muscles and ligaments, and to regulate the heartbeat. Creedite helps address nervous system issues and improves brain function.

ASTROLOGICAL: *None*, **CHAKRA:** *Third Eye*, **ELEMENT:** *Fire*

DIOPSIDE

FAMILY: Inosilicate Pyroxine

CHEMISTRY: $MgCaSi_2O_6$

HARDNESS: 5.5-6.5

STREAK: White

SG: 3.30–3.32

STRUCTURE: Monoclinic Crystal

GEM-QUALITY CHROME DIOPSIDE

DIOPSIDE is a deep green, pyroxine mineral that usually occurs in deep-mantle, igneous rocks brought to the surface by volcanic activity. It has a high magnesium/low silicate composition, and derives its beautiful luster and color from the inclusion of very small amounts of chromium atoms in its structure. It can usually be found with other pyroxenes, such as peridot. Diopside can be found in many locations throughout the world including New York, Russia, South Africa and Burma. It is commercially useful in the ceramics industry and is popular as a semi-precious gemstone for setting in jewelry, even though it is relatively soft. Good quality, uncut specimens can also display beautiful translucence. Other than emerald, very few translucent minerals rival chrome diopside for beauty and rich color.

Diopside is a helpful aid for focusing the mind on learning, especially in a school setting or in new learning situations. In younger children it improves focus, which can help with learning disabilities and dyslexia. It can be used to remove writers block. Diopside is useful when healing from surgery or after a trauma. Women experiencing menopause find it helpful for calming emotions and dealing with physical changes. It has been recommended for use while healing from a heart attack. Diopside is good for healing the lungs from damage caused by smoking and can also be used for detoxification of all sorts.

ASTROLOGICAL: *Gemini*, **CHAKRA:** *Heart*, **ELEMENT:** *Air*

DUMORTIERITE

FAMILY: Boro-Nesosilicates

CHEMISTRY: $Al_7BO_3(SiO_4)_3O_3$

HARDNESS: 7.5–8

STREAK: White

SG: 3.35–3.37

STRUCTURE: Orthorhombic Crystal

DUMORTIERITE

DUMORTIERITE is a very hard, bright-blue nesosilicate, commonly used for ornamentation and carving. It is also used in the manufacture of high grade porcelain. It is sometimes mistaken for sodalite or blue quartz because of its rich color and mottled texture, and it has been used as imitation lapis lazuli. It can easily be distinguished from those by its superior hardness. Dumortierite is usually not uniform in color, but displays characteristic light and dark streaks. Occasionally, iron replaces some of the aluminum in dumortierite, causing it to turn towards pink, red or brown shades. While not too common on the retail or collector's market, dumortierite occurs in many locations around the world including, Madagascar, Canada, Namibia, Norway and Russia. The southwestern United States has also recently become a minor supplier of fine grade dumortierite. These deposits have actually been mined for quite some time, but the dumortierite was erroneously identified as blue quartz.

Dumortierite enhances organizational abilities, self-discipline, and neatness. It is of great use when preparing for long hours of studying or for test preparation. Physically, dumortierite helps with cooling the body and promotes health of the throat, neck and thyroid. Also used for detoxification, calming overstimulation, and for calming hyperactivity. It is an excellent focus stone for children. Dumortierite treats the spleen, the endocrine system, and blood.

ASTROLOGICAL: *Leo,* **CHAKRA:** *Throat,* **ELEMENT:** *Water*

EMERALD

FAMILY: Inosilicate Pyroxine

CHEMISTRY: $Be_3Ai_2(SiO_3)_6$

HARDNESS: 5.5-6.5

STREAK: White

SG: 3.30–3.32

STRUCTURE: Monoclinic Crystal

RAW EMERALD

EMERALD has been a valuable gemstone throughout human history, in widespread use around the world. It is the green variety of the mineral beryl, and is colored by trace amounts of chromium and sometimes vanadium. Emeralds are an extremely hard substance, but because of impurities and inclusion of other minerals, they are prone to fracture. Emeralds without imperfections or inclusions are almost non-existent, so the gemological description, "eye clean" rarely applies to emeralds. Deep green emeralds and other beryls such as heliodor *(yellow)*, morganite *(pink)*, aquamarine *(blue-green)* and goshenite *(clear)* are among the most highly valued and sought-after gemstones in the world. Columbia is the predominant location for mining emeralds today. In Columbia, beryls are found both in limestone beds and in tin and tungsten ore sites. The New England region in the United States is another source of beryls of most types, but very rarely green emeralds.

Emerald is stone of passion, romance and joy. It is said to bring vision and clairvoyance, and can promote deeply spiritual feelings. It symbolizes intelligence, wisdom and truth. Emerald improves brain-function, memory, and communication skills. It is a powerful tool for emotional healing. Physically, it improves disorders of the heart, lungs, spine and muscular system. Emerald has been suggested for use in recovery after infections, and for relieving irritated sinuses and dry eyes. It has a detoxifying effect on the liver and improves diabetes symptoms and fevers.

ASTROLOGICAL: *Aries,* **CHAKRA:** *Heart, Element: Water*

Related entries in this book: Aquamarine, Goshenite, Heliodor, Morganite

EPIDOTE

FAMILY: Sorosilicates

CHEMISTRY: $Ca_2(Al_2Fe_3+)+$
(O,OH,SiO_4,Si_2O_7)

HARDNESS: 6-7

STREAK: Grayish White

SG: 3.30-3.50

STRUCTURE: Orthorhombic Crystal

EPIDOTE IN MATRIX

EPIDOTE is an abundant mineral, being found worldwide. In its finest crystal-line form it displays a rich, deep green color that is translucent to semi-transparent, and is quite valuable as a jewelry gem. Epidote also displays very strong pleochromism, which means its apparent color is dependent on the viewing angle. It can appear green, black or yellow depending on the viewer's position in relation to the lighting source. In massive form, the shades of green can be so dark as to appear almost black. It has a complicated molecular structure, containing iron, calcium, aluminum, silicon, oxygen and hydrogen. Epidote usually occurs in marble and other metamorphic rocks. It can also be found as the alteration product of hydrothermal contact in igneous rocks, and is often found in association with feldspar, garnet, mica and other minerals. Epidote is responsible for the green patches in the pink and green colored mineral unakite *(the pink is orthoclase feldspar)*.

Epidote is an excellent stone for tuning-in to nature in general, and more specifically, to one's immediate environment. It has a calming, grounding and relaxing effect on the bearer. Its affinity to nature is particularly useful for those living in a crowded and noisy urban setting, as it can remind us that we always have an underlying connection to the Earth. Epidote is a stimulant and can boost the immune system. It is also a strong ally for assisting with the healing process after trauma, surgery or long-term illness.

ASTROLOGICAL: *Gemini*, **CHAKRA**: *Gemini*, **ELEMENT:** *Earth*

GEM QUALITY EPIDOTE CRYSTAL

FELDSPAR

FAMILY: Tectosilicates

HARDNESS: 6.0 (Defining)

STREAK: White

SG: 2.55–2.63

STRUCTURE: Orthorhombic Crystal

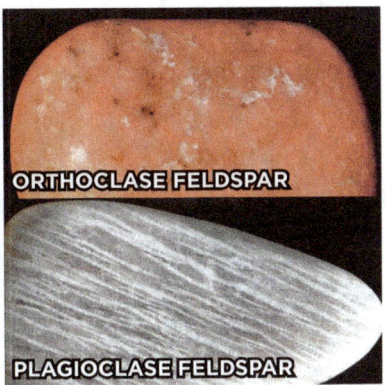

ORTHOCLASE FELDSPAR

PLAGIOCLASE FELDSPAR

FELDSPAR The feldspar silicate group is the most widespread family of minerals on the Earth. It is estimated that feldspars make up over 60% of the rocks in the Earth's crust. Orthoclase is specifically a potassium-rich variety of feldspar and is mineralogically referred to as K-feldspar. The other major group of feldspars are either sodium or calcium-rich, and are referred to as plagioclase feldspars. Orthoclase is usually pink but also occurs in yellow, gray or white. Pink orthoclase feldspar is responsible for the pink areas present in the bi-colored mineral unakite *(the green is epidote)* and is also present as the pink flecks in most granite rocks. Feldspars typically form from silicate magmas and are also found in a wide range of metamorphic rocks. Feldspar has many important commercial uses, as it is widely used in the production of glass, ceramics and industrial abrasives.

ORTHOCLASE FELDSPAR *(KAlSi3O8)* encourages cooperation among individuals, and offers insight into group experiences. It can help with finding creative ways to obtain personal goals. It eases the psychological effects of aging, and lessens the fear of getting old. Orthoclase is also beneficial for house and garden plants because of its deep and widespread connection to the Earth.

Related entries in this book: Moonstone

PLAGIOCLASE FELDSPAR *(NaAlSi3O8 or CaAl2Si2O8)* is a stone of light, compassion and understanding. It is useful when exploring spirituality and mysticism, as it promotes serious and focused reflection into new revelations, before revealing higher knowledge. Physically, plagioclase soothes irritability, headaches and nausea.

Related entries in this book: Moonstone, Labradorite, Sunstone

ASTROLOGICAL: *Aquarius*, **CHAKRA:** *All*, **ELEMENT:** *Earth*

FLUORITE

FLUORITE AND GALENA

FLUORITE

FAMILY: Halides

CHEMISTRY: CaF_2

HARDNESS: 4.0 (Defining)

SG: 3.18–3.21

STRUCTURE: Isometric Crystal

BLUE FLUORITE

FLUORITE can be found in a variety of bright, transparent colors, and also as colorless crystal forms of optical quality *(spar)*. It is known for glowing brilliantly in short-wavelength UV light, lending its name to the term "fluorescent". Fluorite typically forms at shallow depth in the Earths crust, and in relatively cool conditions, where fluorine-rich fluids can permeate calcium-rich sedimentary rocks, such as limestone. It is also commonly found in association with metal ore veins of silver and lead *(galena)*. Fluorite is relatively soft and brittle, and cleaves easily along 4 crystal planes, making it difficult to cut or polish for jewelry purposes. Unlike many translucent minerals, the coloring of fluorite does not come strictly from chemical impurities, but also partly from micro-fractures in the interior of the crystal which refract some colors of light better than others. Fluorite is the official mineral of the State of Illinois but is not mined there much commercially anymore. Fluorite is extremely common and can be found almost worldwide.

BLUE FLUORITE calms emotions and stimulates clear communication. It alleviates irritation of the nose, throat and ears. Same as all other fluorites, blue fluorite is excellent for the health of teeth and gums.

ASTROLOGICAL: *Capricorn,* **CHAKRA:** *Heart,* **ELEMENT:** *Water*

GREEN FLUORITE

PURPLE FLUORITE

GREEN FLUORITE brings the body's hormones into balance and recharges the chakras. It enhances the immune system, boosts cell regeneration and helps with assimilation of nutrients. Green fluorite heals scars, both emotional and physical. It can help with sore throat and stomach ailments such as ulcers and indigestion. It is suggested for use easing the pain and stiffness of arthritis. Green fluorite is used to bring relief from long-term insomnia.

ASTROLOGICAL: *Scorpio,* **CHAKRA:** *Sacral,* **ELEMENT:** *Water*

PURPLE FLUORITE increases visions and brings clearer awareness of spiritual contact. It aids with balancing the body and spirit. It is a very useful stone when preparing for meditation or spiritual journeys. It helps connect one's intuition to the rational mind, and to create a natural balance between the two. Purple fluorite promotes healthy skin, teeth, bones and muscles.

ASTROLOGICAL: *Virgo,* **CHAKRA:** *Third Eye,* **ELEMENT:** *Air*

RAINBOW FLUORITE is useful for improving clarity and focus. It brings serenity and feelings of contentment with yourself and your surroundings. Healers use rainbow fluorite for detoxification and for cleansing of the lungs, throat and digestive systems. Rainbow fluorite is also a symbol of happiness and joy.

ASTROLOGICAL: *Leo,* **CHAKRA:** *Crown,* **ELEMENT:** *Water*

RAINBOW FLUORITE

FOSSILS

FOSSILS

AMMONITE FOSSIL

FOSSILS would normally be addressed in a book about paleontology, or paleo-biology, but there are several reasons to include them in this book. Geologically, fossils can be important indicators of the history of a rock formation, its age and whether the rock was part of land or sea at the time of deposition. For organic matter to truly be considered a fossil and not simply preserved, it has to have been chemically replaced by inorganic mineral material, actually becoming solid rock. This process can take thousands to millions of years. Most fossils are com-posed of some sort of cryptocrystalline quartz in the form of opal or chalcedony. Fossils have held intrigue throughout human history because of their mystery and allure, and the same holds true today; collectors still have a strong fascina-tion for fossils. The collector's market remains vigorous and widespread, making fossils abundantly available and modestly priced.

Some cultures and traditions have developed specific healing uses and meta-physical properties for many fossil types, however for the purposes of this book, we will only describe the one clear attribute that all fossils have in common. Because of their incredible age and their appearance in almost every place on Earth, fossils are used by past-life seekers for exploring knowledge from past experience. Fossils represent stability, purpose, and a greater awareness of the connection between the Earth and all of its life forms.

BRACHIOPOD FOSSIL

AMMONITE FOSSIL Ammonites are some of the most easily recognized fossils throughout the world. Their sheer abundance, and striking spiral shape, makes them a favorite with collectors of all sorts. These mollusks lived in oceans around the globe between approximately 400 to 66 million years ago and are closely related to nautilus species in our oceans today . Not only is this a vast period of time, but because they have been studied so closely, their evolving shapes and structures provide information about the age of any geologic layer in which they can be found. Fossils such as these are be called, "index fossils" and are some of the best tools that geologists can use to date rock formations. Many ammonite fossils display wonderful opalescent rainbow colors on their surfaces. Ammonites are frequently cut in halves and polished, to display the chambered spiral shapes of the inner-shell.

BRACHIOPOD FOSSIL Brachiopod fossils are found in rock deposits representing a vast range of time. They can be found in regions where sea water was once shallow and calm. Their bodies seem to be similar in structure to modern bivalves, and there are even some somewhat similar species of brachiopoda currently living in our oceans. A good illustration of how much change the Earth has gone through over hundreds of millions of years is the fact the brachiopod is the Official State Fossil of Kansas, in the heartland of the USA, more than 1000 miles from the nearest ocean! Brachiopod fossils can be found intact, weathered out of their typical limestone beds, and showing all surfaces clearly. Our friends in the Upper Peninsula of Michigan showed us where to collect excellent specimens from between limestone layers, only a few feet from the shore of Lake Michigan.

CHRINOID FOSSILS

COPROLITE

CHRINOID FOSSIL Crinoids are fossil remains of an animal closely resembling modern sea lilies, and should not be mistaken for plant fossils. Their stems are the most commonly found part of these ancient creatures, likely due to their toughness, as opposed to the fleshy "feathers" and mouths of their upper bodies, or their small roots. They are often confused with other fossils, such as the spinal columns of vertebrates, or fish bones. Crinoids date back as far as the Ordovician Era *(485-445 million years ago)*. They can be found in many limestones worldwide, and abundantly in Western Illinois, along the Mississippi River Valley.

COPROLITE fossils being a smile and a giggle to most collectors who encounter the, for the first time. A coprolite specimen is fossilized animal dung. There is a variety of shapes, sizes, colors, textures, depending on what animal left the specimen for future collectors. Many people refer to coprolite as fossilized dinosaur poop, but the fact is, most specimens collected came from any of several species of ancient lizards. We have specimens in our collection ranging in size from one to six inches. Museums have specimens weighing several hundred pounds! Having some coprolite specimens in the collection bring a smile to viewers and are an excellent conversation starter.

ORTHOCERAS FOSSILS

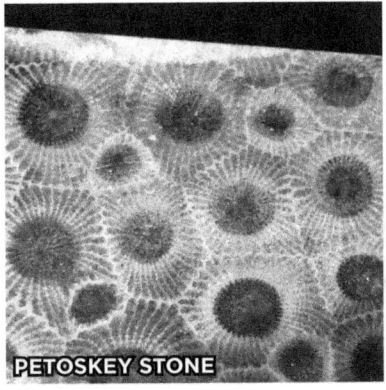

PETOSKEY STONE

ORTHOCERAS FOSSIL Orthoceras fossils are very popular with collectors due to their striking features and also because they are readily available in the marketplace. These ancient cephalopods lived during the ordovician period, over 450 million years ago, in a region that is now the Baltic Sea. For collectors, the tapered, segmented, tubular fossil bodies, are purposely half-exposed in their unusually black, limestone matrix, which further accentuates their features which appear white or gray. Many display specimens have multiple orthoceras bodies crowded in a very small space, indicating the possibility that either tidal forces or mating-swarms gathered numerous creatures into a tight space before they died suddenly, and were preserved. Mudslides in shallow waters could be responsible for so many creatures perishing and being subsequently buried in a low-oxygen situation needed to interfere with decomposition, this preserving the body for fossilization. These fossils are quite inexpensive and readily available.

PETOSKEY STONE. These beautifully patterned fossil stones are remains of shallow-sea corals, that lived off of what is now the Lake Michigan shore of Northwest Michigan. Between the towns of Petoskey and Traverse City, they are found directly on shore or in the shallow water, particularly after storms and wave action bring them from deeper waters. The coral bed was broken-up long ago by repeated glaciation in the region, and later submersion in the lake, where wave action has tumble-polished many to palm or pebble-sized specimens. The coral patterns on the surface are obvious as they resemble many existing corals. The stones take on a nice shine when polished, and much jewelry and adornment has been made with Petoskey Stone fossil specimens.

STROMATOLITE

TUMBLED STROMATOLITE

STROMATOLITE is the term for a living or fossilized microbial mat *(layers)* of cyanobacteria *(similar to blue-green algae)*. These colonies may reach up to a meter or more in height, and tend to be found mostly submerged, in shallow water. Fossilized stromatolites exhibit striking ringed and banded patterns. Unlike most fossils which are typically silicate opal in chemistry, stromatolite fossils are sedimentary rocks. Geologic records of ancient stromatolites similar in structure to living ones, have been dated to be at least 1.3 billion years old. Living stromatolite colonies can be seen in Australia, Brazil, and Chile and elsewhere.

RAW STROMATOLITE

FULGARITE

FAMILY: Lechitelierite (Silica Glass)

CHEMISTRY: SiO_2
(Fused by Lightning Strike)

HARDNESS: 6.5

STREAK: White, None

SG: 2.50–2.71

STRUCTURE: Amorphous Rock

FULGARITE

FULGARITE is classified as a variety of the mineraloid lechatelierite, which is a glass composed primarily of silica from sand. Fulgarites are naturally hollow, glass tubes formed when lightning fuses sand on the ground at the point of the strike of electricity. The high temperature instantaneously melts the silica, which then cools and crystallizes into glass so quickly, that the generally rod-like or tubular shape of the electric discharge from the lightning is retained. Similar mineraloids can be formed by meteor impact or in industrial settings where high-voltage electricity comes in contact with sand. Fulgarites are quite rare, and are mainly found on mountaintops or other areas where lightning strikes are common. Large deserts like the Sahara, are common places for collectors to hunt for fulgarites, and while there may be many millions of them, in such a vast desert, they are extremely difficult to find. Mineral and rock shops usually have fulgarites for sale at modest prices. For your own collection, a specimen of a rock "made from a lightning strike" is a unique and interesting piece, and a real conversation starter. Decent specimens are commonly sold at mineral and rock shops in small display boxes with printed descriptions, for gift purposes. Quite nice!

Fulgurite enhances communication on the physical plane, and strengthens connections to spiritual realms and to entities in those realms. Fulgarite has a strong energy that enhances focus and concentration, and improves brain function in general.

ASTROLOGICAL: *Virgo,* **CHAKRA:** *Third Eye,* **ELEMENT:** *None*

FUCHSITE (Chrome Muscovite)

FAMILY: Phyllosilicates

CHEMISTRY: $K(Al,Cr)_3 Si_3 O_{10} (OH)_2$

HARDNESS: 2-3

STREAK: White

SG: 2.85-2.88

STRUCTURE: Monoclinic Crystal

FUCHSITE

FUCHSITE (CHROME MUSCOVITE) is a chromium-rich form of muscovite (mica). It is also commonly spelled "fuchite". It is far less flaky and cleavable than muscovite mica. This trait makes it possible for some fuchsite to be tumbled and polished or cut for jewelry and decoration purposes. Fuchsite is mined in Sweden, Austria, Brazil and the USA. Like many phyllosilicates, the layering of its varied mineral inclusions can present a stunning shimmer. High-quality, polished examples show a bright chatoyance, or cat's eye effect. We have specimens of polished and raw fuchsite in our family collection, and both make wonderful display pieces due to their bright coloring and shine. Fuchsite is frequently found in a popular rock-combination with ruby, and is widespread commercially. "Ruby in fuchsite", or simply, "ruby fuchsite", has become very popular with collectors and healers, and is abundant and inexpensive in the retail market. It is also available in a wide variety of carved shapes and small statuary pieces.

Fuchsite is used for relieving stress in general. It helps the bearer to abandon lost causes and halt the tendency to help others before taking care of one's self. Fuchsite helps to overcome codependency issues in relationships and helps recovery from trauma and physical injury. Fuchsite is recommended by healers to eliminate back pain and to loosen back muscles, making the body more balanced and flexible. It treats repetitive motion issues such as carpal tunnel syndrome, so it is useful for frequent computer users. Fuchsite is good for the circulatory system and the production of red and white blood cells.

ASTROLOGICAL: *Aquarius,* **CHAKRA:** *Solar Plexus,* **ELEMENT:** *Earth*

Related entries in this book: Lepidolite, Muscovite, Ruby in Fuchsite

GALENA

FAMILY: Sulfides

CHEMISTRY: PbS

HARDNESS: 2.5

STREAK: Gray

SG: 7.22–7.81

STRUCTURE: Isometric Crystal

GALENA

GALENA is an abundant sulfide mineral and the major ore of lead. It can be found worldwide, most often in conjunction with silver or zinc deposits. It can occasionally be found in association with calcite, fluorite and barite. It has a very bright, shiny, silver-gray appearance. Typically, it forms in large masses but most commercially mined pieces are fractured, and show beautiful staggered, cleavage surfaces which increase its stunningly reflective appearance. It was mined extensively in Illinois for its lead content until the 1990's. Like most of the heavy metal minerals, galena is very dense and very soft and pliable. It is not safe to handle frequently because of its lead content. It should not be handled by children unless supervised, and hands should always be washed after touching galena or any form of lead.

Galena balances the male and female aspects of personal energy and leads to a wholeness and completeness of the self. While galena is not very frequently found in the crystal healing arsenal, it is considered a powerful stone and is a good remedy for blockages that limit personal growth. Galena is useful for improving conditions of the blood and circulatory system as well as being beneficial to the lungs. It can help with chest congestion and eases the symptoms of pneumonia.

ASTROLOGICAL: *Scorpio*, **CHAKRA:** *Base*, **ELEMENT:** *Earth*

GARNET

FAMILY: Nesosilicates

CHEMISTRY: $X_3Y_2(SiO_4)_3$

HARDNESS: 3.70–4.10

STREAK: White

SG: 6.5–7 .5

STRUCTURE: Tetragonal Crystal

ALMANDINE GARNET

GARNETS are a loosely associated group of hard and colorful, gem-quality silicate minerals. In addition to silicate molecules, their chemical structure can include calcium, magnesium or iron, in combination with aluminum, iron or chromium. Garnets have been used as gem stones for several millennia and are still precious and popular today. They are formed in high-heat, high-pressure zones at great depths in the Earth, and brought to the surface either through tectonic uplift or through volcanic activity. Garnets are found in several locations, most famously China, Myanmar, South Africa, Sri Lanka, India, and the U.S.A. The colors and variations of garnets are extensive, and include: almandine *(orange-red)*, grossularite *(green)*, pyrope *(wine-red)*, spessartine *(orange)*, hessonite *(brown or orange)*, demantoid *(green)*, topazolite *(yellow)*, melanite *(black)*, and rhodonite *(pink)*. Crushed garnet, both natural and synthetic, is used extensively as an abrasive such as sandpaper. Raw and uncut garnets, such as some in our family collection, are readily available at retail shops for collectors and healers, and are not prohibitively costly. For me, this is a blessing because I absolutely love all varieties garnets!

Garnet activates and energizes all chakras. It revitalizes, purifies and balances spiritual energy and brings either serenity or passion as appropriate. It is a stone of love and devotion, and sharpens perceptions about one's self and others. Garnet brings courage and self-confidence. Physically, garnet rejuvenates the body and stimulates the metabolism. It is used to treat disorders of the spine and skeletal system. It is thought to be helpful for repairing damaged cellular structure and regenerating DNA. Garnet protects the heart, lungs and blood and is a boost for the immune system.

ASTROLOGICAL: *Aries,* **CHAKRA:** *All,* **ELEMENT:** *Fire*

PYROPE GARNET

SPESSARTINE GARNET

There are two main subgroups of garnet minerals; Ugrandite *(calcium garnets)* and Pyralspite garnets *(calcium-free)*.

THE PYRALSPITE FAMILY OF GARNETS These garnets are the calcium-free garnets. They are ideochromatic, meaning their specific colorings are due to their basic chemistry, not from minor impurities *(allochromatic)*.

Pyralspite garnets include almandine *(orange-red)*, pyrope *(wine-red)* and spessartine *(orange)*. The varieties of garnets shown in the photos are from our family collection and include raw and polished specimens of these gemstones.

THE UGRANDITE FAMILY OF GARNETS These garnets all contain calcium in their basic molecular structure. They are considered allochromatic, meaning other chemical impurities give them their definitive colorings.

The ugrandite garnets include uvarovite *(green)*, andradite *(yellow green)* demantoid andradite *(light green)*, melanite andradite *(black)*, and grossular. Grossular garnet varieties include hessonite *(orange)*, rosalite *(pink)*, leuco *(clear)*, and tzavorite *(green)*.

UVAROVITE GARNET

ANDRADITE GARNET

DEMANTOID GARNETS

MELANITE GARNET

GROSSULAR GARNET

HESSONITE GARNET

TSAVORITE GARNET

GEODES

GEODES

GEODE

GEODES are hollow nodules of stone that display interior growth of crystalline or banded minerals. They can be found in many locations worldwide, usually in sedimentary or igneous formations. Theory states that hollow pockets form in lava flows and are later subjected to saturation from mineral-rich liquids, which stimulate the growth of crystals. Likewise, in sedimentary deposits, plant or organic materials can decay, leaving voids in the rock that later fills with crystals in much the same way. Some of the best crystals available are found in geodes, since there is no interference with perfect crystal growth. Some geodes found in Brazil are large enough for a person to walk in! Geodes are popular with collectors and are readily available at stores and gift shops. We recently had the good fortune to mine for our own geodes in Hamilton Illinois/Keokuk, Iowa. This area near both banks of the Mississippi River, is famed for geode hunting and hosts conventions and tours. Since we found wonderful and quite numerous specimens of our own, in a relatively short period of time, we whole-heartedly agree with the title, "Geode Capital of the World"! We brought back to our family collection a multitude of excellent specimens with a huge variety of minerals present. The experience of working the deposits with our young boys, prying these "gems" out of the rock wall with picks and chisels, and cracking them open to reveal the treasures inside, was a thrilling experience for all of us.

Geodes will generally have the same spiritual and healing properties as the minerals that are present in the specimen. Geodes also promote communication in groups, a sense of community, and an understanding of the connection between all living things. They represent the circle of life.

GOLDSTONE
(Aventurine Glass)

FAMILY: Silicate Glass
(with Copper Oxide)

CHEMISTRY: $SiO_2 + (CuO_2)$

HARDNESS: 5.5 (glass)

STREAK: None

SG: 2.50–2.80

STRUCTURE: Amorphous Glass

GOLDSTONE

GOLDSTONE (AVENTURINE GLASS) is a manufactured glass, but we have included it in this book because it is a very attractive and popular item with collectors and even more popular with users of healing crystals. This beautiful, glittering glass is made with very pure silica sand, and has copper flakes added for sparkle and shine. Production takes place in a low-oxygen atmosphere which prevents tarnish and oxidation, and helps with even dispersion of the metal flakes within the glass. Goldstone is produced in many colors, dark blue or golden-red being the most popular. It is very bright, cuts or carves well for jewelry, and rivals some semi-precious gemstones for luster and beauty.

Goldstone is said to be a stone of ambition, giving the bearer creative inspiration and the energy to complete projects. It enhances strength, confidence and courage, and is a reminder to keep a positive attitude when faced with difficult challenges or obstacles. Being a powerfully uplifting stone, goldstone will also elevate the mood and energy of those around the user. This makes it an excellent stone to use for social settings and parties. Physically, goldstone is used by healers to help with stomach and digestive problems. It is also soothing to arthritis sufferers, and is generally helpful for conditions of the bones and joints as well as circulatory function.

Related entries in this book: Obsidian Glass, Opalite

GOSHENITE

FAMILY: Cyclosilicates

CHEMISTRY: $Be_3Al_2(SiO_3)_6$

HARDNESS: 7.5–8.0

STREAK: White

SG: 2.74–2.76

STRUCTURE: Hexagonal Crystal

GOSHENITE

GOSHENITE is the white or colorless variety of beryl, from the same crystal family as the more popular and colorful emerald and aquamarine. It used to be assumed that goshenite was a pure form of beryl, with no impurities to create color as in other precious beryls. However with modern detection techniques, trace impurities of iron, titanium, vanadium, scandium and a handful of other elements have been found. Using irradiation treatment, it is possible to bring out the color of these impurities and create attractive gemstones. Translucent, gem-quality goshenite crystals are fairly rare, and quite costly. This colorless stone has never become very popular in the jewelry trade. Even opaque crystals of this beryl can be difficult for collectors to find.

Goshenite is high vibration stone used by energy workers to promote clear thinking in both the healer and the subject. Because of its high energy, it is used as a spiritual booster as well as being a very strong, emotional healing crystal. It is said to be useful physically, for removing infections and viruses from the body. Goshenite is not very common in healing markets so it is usually best to scour the mineral collector's resources.

ASTROLOGICAL: *Scorpio*, **CHAKRA:** *Crown*, **ELEMENT:** *Water*

Related entries in this book: Aquamarine, Emerald, Heliodor, Morganite

GYPSUM DESERT ROSE

FAMILY: Sulfides

CHEMISTRY: $CaSO_4\ 2H_2O$

HARDNESS: 2.0 (Defining)

STREAK: White

SG: 2.31–2.33

STRUCTURE: Monoclinic Crystal

GYPSUM DESERT ROSES

GYPSUM (SELENITE) DESERT ROSE is a unique and beautiful crystal structure that can form individually or in large, connected groupings. The unique and intricate, bladed crystals of a desert rose form in shallow, sandy environments where gypsum has precipitated from water through sand, forming crystals around the sand grains. Upon weathering away of surrounding sand grains, the desert rose emerges on the surface. Desert rose is a descriptive name given to other rose-shaped crystalline formations other than gypsum, such as barite. A gypsum desert rose is slightly harder and denser than the simple, massive form gypsum, probably due to its content of silicate sand. However, silica rarely entirely replaces *(pseudomorphs)* the gypsum, as it can do in some barite crystal formations. The gypsum and the silicate sand are bound together as a matrix, making this a rock, not a mineral.

Desert rose is said to bring increased mental sharpness, and clarity of thought, as well as improved perception and intuition. Desert rose is used in crystal healing to dispel worries and increase focus during times of distraction and disruption. This makes it an excellent tool when preparing for meditation or yoga, or when preparing for a test or school exam. Physically, the desert rose promotes proper alignment of the spine and skeleton, relieves stress, and encourages repair of damaged body tissues.

ASTROLOGICAL: *Gemini,* **CHAKRA:** *Heart,* **ELEMENT:** *Air*

Related entry in this book: Selenite

GYPSUM DESERT ROSE

HALITE

FAMILY: Halides

CHEMISTRY: NaCl

HARDNESS: 1.5–2.5

STREAK: White

SG: 2.11–2.16

STRUCTURE: Isometric Crystal

HALITE is also known as rock salt, table salt, or sodium chloride. It is probably the most commonly used mineral in our daily lives. In fact, it is so common that it is often overlooked and under-appreciated as a mineral, which is why we decided to include it in this book. Everyone has salt in their home for daily dietary use or as a bath additive or for melting ice, preserving foods, or even for making ice cream. But halite is also available to collectors and healers in a variety of attractive crystalline forms. Being an essential mineral, we feel it deserves a proud place in the family mineral collection. Recently, imports of Himalayan salt have become very popular, and because of its beautiful, gentle, orange hue and its slight translucence, it is often carved into figurines or hollowed-out and used as electric lamps that give a beautiful glow. Halite commonly occurs as an evaporate mineral layer, where a lake or sea-bed has dried-up, but because it is highly soluble in water, these environments must remain dry for long periods of time for good crystals to grow. Many of these salt beds are found below the surface, protected from moisture. Some of these underground salt beds may be up to hundreds of feet thick and cover vast areas. Halite has been mined and used by almost every civilization known, for thousands of years. In some regions, in the past, salt was a precious natural commodity even more valuable than gold.

Halite encourages us to spread good will and lend aid and support to others. It discourages negative thoughts, and thus elevates moods and promotes happiness. Halite also builds determination to accomplish goals and reminds the bearer of the importance of independent and original thought. Physically, halite is recommended for detoxification and purification of the mind and body. It is also used as a crystal for purifying the air and the immediate environment, and is recommended for people suffering from allergies and sinus issues.

ASTROLOGICAL: *None,* **CHAKRA:** *Heart,* **ELEMENT:** *Air*

HELIODOR

FAMILY: Cycloilicates

CHEMISTRY: $Be_3Al_2(SiO_3)_6$

HARDNESS: 7.5–8

STREAK: White

SG: 2.276

STRUCTURE: Hexagonal Crystal

HELIODOR

HELIODOR is golden-yellow beryl, from the same crystal family as the more popular emerald and aquamarine. While a perfectly clear, inclusion-free emerald is practically non-existent, heliodor is known for superior clarity and purity of color. Heliodor can be found most often with other colors of beryl in granite pegmatite formations worldwide, most notably the United States, Europe and Russia. Recently, heliodors have made it to the commercial market from locations in the Middle-East, but reportedly they are actually aquamarines that have been irradiated to the golden beryl color. Like emeralds, heliodor is a precious gem stone and is priced accordingly. However, less costly, non-gem, specimens are occasionally found at rock shops or from sources online. Even in lower quality crystals, the deep, rich color makes it obvious why heliodor is treasured as a gemstone.

Heliodor is a symbol of self-reliance and prosperity, thus it is of use to the self-employed or those undertaking new business ventures. Heliodor is also good for focus and memory, making it and excellent study stone. Physical uses may include promoting proper health of the digestive system, and speeding recovery from surgeries.

ASTROLOGICAL: *Leo*, **CHAKRA:** *Solar Plexus*, **ELEMENT:** *Water*

Related entries in this book: Aquamarine, Emerald, Goshenite, Morganite

HELIOTROPE (Bloodstone)

FAMILY: Cryptocrystalline Silicates

CHEMISTRY: SiO_2

HARDNESS: 6.5–7

STREAK: White

SG: 2.81–2.88

STRUCTURE: Trigonal Hexagon Crystal

HELIOTROPE

HELIOTROPE (BLOODSTONE), also known as bloodstone, is a very dark, green variety of quartz chalcedony. Its blood-red inclusions are due to iron oxide impurities in the form of opaque, red jasper. The green coloring is from chemical inclusion of chlorite in the molecular structure. In some cases the inclusions may be yellow, in which case the mineral is given the name plasma, but this is a very uncommon and a fairly rare stone. Heliotrope has historically been mined primarily in India but the supply may be dwindling. There are still some lesser sources of the stone, including Australia, Scotland, Brazil and China. Almost all heliotrope currently brought to the market is in tumbled and polished form, which accentuates its deep, dark coloration. When raw stones are found, they are usually small and of inferior color and quality.

Bloodstone brings intuition, increases creativity, and provides solid grounding. It repels negative environmental energy, helping to overcome unhealthy exposure to toxins, radiation and EM energy fields. Bloodstone stimulates vivid dreams and is a powerful rejuvenator from exhaustion. Physically, it purifies the blood and helps to generate new blood cells. It detoxifies the liver, intestines, kidneys, spleen and bladder. Bloodstone improves and supports blood circulation and heals infection and pneumonia. It is a potent ally when fighting inflammation of any part of the body.

ASTROLOGICAL: *Aries,* **CHAKRA:** *Base,* **ELEMENT:** *Fire*

HEMATITE

FAMILY: Oxides

CHEMISTRY: Fe_2O_3

HARDNESS: 5.5-6.5

STREAK: Red

SG: 5.30-5.33

STRUCTURE: Rhombohedral Crystal

BOTRYOIDAL HEMATITE

HEMATITE is a common iron oxide and is mined as the main source of iron ore. It is very heavy and dense, and displays a nice metallic shine. It occurs primarily in deposits where there once would have been standing water or a hydrothermal source. Iron precipitates readily out of water and forms as banded, solid layers. Iron can also be present in igneous rock formations due to volcanic activity moving iron towards the Earth's surface from great depths and combining with atmospheric oxygen. For mineral collectors there is a wide variety of hematite forms from which to choose. One popular and attractive type of massive specimen that we chose for our family collection is "botryoidal" hematite, which describes the groupings of small grape and pea-sized bumps. Because of its unusual visual appeal, its nice metallic luster and its weight, it attracts a lot of attention from guests when viewing our collection. Our family visited an abandoned iron-ore mine in Vulcan, Michigan and we were able to collect several large ore specimens there that are also part of our collection. Some smaller samples polished-up nicely and made excellent pieces for jewelry.

For healers, hematite is considered a stone of great strength and stamina. It has deep roots in the Earth and connects us to the planet and its energies. Physically, it brings health to the circulatory system and treats blood conditions such as anaemia. It improves kidneys function and is recommended for tissue regeneration. Hematite promotes absorption of iron into the body and the formation of new blood cells. It is recommended for easing leg cramps, anxiety and insomnia. It also helps improve spinal alignment and is said to speed recovery from broken or fractured bones.

ASTROLOGICAL: *Aries*, **CHAKRA:** *Base*, **ELEMENT:** *Fire*

HERKIMER DIAMOND

FAMILY: Quartz Silicates

CHEMISTRY: SiO_2

HARDNESS: 7

STREAK: White

SG: 2.71–2.73

STRUCTURE: Trigonal Hexagon Crystal

HERKIMER DIAMOND

HERKIMER DIAMOND is not a diamond at all, but rather a near-perfect, double-terminated quartz crystal famously found in mines in Herkimer, New York, U.S.A. These unique crystals are found in pockets, or "vugs" in the local dolostone *(dolomitic limestone)* formations. They are believed to be over 500 million years old, and formed when vegetation, trapped in pockets in sediment layers, was later replaced with silica either from bacterial decomposition, or by intrusion of silica-rich water. Herkimer diamonds are sold on the market for jewelry, or as crystals for collectors and healers. They can be mined from private properties in New York, for a fee. We have an example in our family collection that has several "diamonds" still in their native dolomite matrix. These types of specimens are excellent for display.

Herkimer diamond is a perfect stone to use for attaining higher visions, for guidance during astral travel, and to amplify the power of other stones. Physically, Herkimer diamonds promote deep and restful sleep, and boost the body's immune system.

ASTROLOGICAL: *Libra,* **CHAKRA:** *Crown,* **ELEMENT:** *Air*

Related entry in this book: Quartz

HIDDENITE

FAMILY: Oxides

CHEMISTRY: $LiAl(SiO_3)_2$

HARDNESS: 5.5–6.5

STREAK: Red

SG: 5.30–5.33

STRUCTURE: Rhombohedral Crystal

HIDDENITE

HIDDENITE is a spodumene mineral related identically to kunzite, except kunzite typically displays a pinkish hue as opposed to hiddenite's pale green translucence. It is named for William Earl Hidden who was credited as one of the discoverers of the mineral. In fact, he may not have discovered it, but was instead responsible for sending-out the first specimens for chemical identification, receiving confirmation that this was a new variety of spodumene. Hiddenite is fairly soft and brittle, so it is difficult to use for jewelry or decorative purposes. When fine examples are faceted, they display beautiful clarity and light. Some raw specimens actually have decent visual appeal and are quite nice in the mineral collection.

Hiddenite is a strong heart chakra stone, which helps when building personal relationships that last and grow closer with time. It is a benefit to the alignment of the chakras and regenerating one's aura. Physically, hiddenite is useful for boosting energy and providing clarity of thought and sound thinking.

ASTROLOGICAL: *Scorpio,* **CHAKRA:** *Heart,* **ELEMENT:** *Air*

HOWLITE

FAMILY: Inosilicates

CHEMISTRY: $Ca_2B_5SiO_9(OH)_5$

HARDNESS: 3.5

STREAK: White

SG: 2.53–2.59

STRUCTURE: Monoclinic Crystal

WHITE HOWLITE

HOWLITE was discovered in 1868 by Henry How, a geologist and mineralogist. It was at the time considered undesirable waste *(gangue)* in gypsum deposits being mined there. Currently most howlite is mined in California and occurs in deposits with other borates and evaporite minerals. It typically forms as lumpy nodules and ground-masses, with actual crystals being somewhat rare. Inclusions of thin, black streaks are commonly interlaced throughout the nodules, making a very attractive display specimen. It is a soft but fairly durable rock, and having no natural cleavage, it is resistant to cracking. It takes a polish well, making it common as a tumbled stone sold in shops. Howlite is found naturally only in shades of white and gray, however dyed samples are quite common. Howlite is often dyed and sold as counterfeit turquoise, which is far more precious and valuable. Close examination of its texture and lack of hardness can help to distinguish a fake.

Howlite is a gentle healing stone that soothes emotions, slows overactive minds and helps calm hyperactivity. It is a great aid for achieving deep and restful sleep. Families with children will find this a valuable tool to have on hand, and a small howlite stone can be placed under a child's pillow before bedtime. Howlite can also be used to facilitate deeper self-awareness, promoting better personal expression and confidence. Howlite is a trusted assistant for eliminating aches and pains and for reducing muscle strain and stress. It also helps the body absorb calcium into the bones and teeth.

ASTROLOGICAL: *Gemini,* **CHAKRA:** *All,* **ELEMENT:** *Air*

HYPERSTHENE

FAMILY: Pyroxine Inosilicates

CHEMISTRY: $(Mg,Fe)SiO_3$

HARDNESS: 5.5

STREAK: White

SG: 3.10-3.33

STRUCTURE: Orthorhombic Crystal

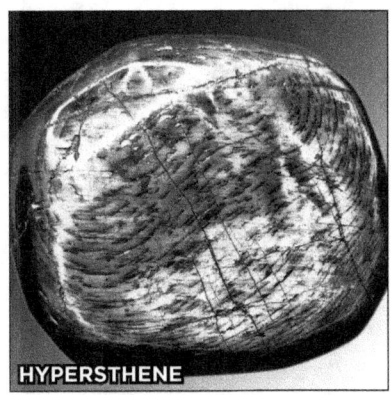
HYPERSTHENE

HYPERSTHENE forms in either igneous or metamorphic rocks, and has also been found as a trace mineral in some iron meteorites. It is a member of the enstatite and ferosilite series of minerals. Enstatite is a magnesium pyroxine mineral that contains little or no iron. Ferosilite contains high concentrations of iron but little or no magnesium. Hypersthene has almost an equal amount of both elements. With a higher ratio of iron to magnesium the mineral is called bronzite. Hypersthene displays a remarkable pleiochromism *(optical effect of changing its apparent color when viewed or lighted from different angles)*. The typical color shift ranges from dull blackish-brown to a shiny-bronze and usually appears in distinct glowing bands which give a sense of depth to the stone. The main source of high quality hypersthene is the Island of Paul off the Northeast coast of Labrador. It is also currently mined in locations as diverse as Australia and upstate New York. It is a stunning and beautiful stone and is at the top of the list of our family favorites.

Hypersthene promotes deep and clear meditational states. It also brings a clear sense of personal limitations and a sense of moral obligation. In this respect, it helps one to be aware that positive energy output can bring success to social and business endeavors. It is helpful for treating sleep disorders and ending insomnia. Hypersthene is very good at releasing tension and stress, providing relaxation or energization, whichever is needed. It helps calm children's fear of thunderstorms and lightning. Hypersthene tends to be slightly more costly than other healing crystals, and a little scarce to find. Once found, it often becomes a very powerful aid to which the healer will return to frequently.

ASTROLOGICAL: *Libra,* **CHAKRA:** *Third Eye,* **ELEMENT:** *All*

Related entry in this book: Bronzite

IOLITE (Cordierite)

FAMILY: Silicates

CHEMISTRY: $(Mg,Fe)_2 Al_3 (Si_5AlO_{18})$

HARDNESS: 7.0–7.5

STREAK: White

SG: 2.60–2.66

STRUCTURE: Orthorhombic Crystal

IOLITE (IN WARM LIGHT)

IOLITE (CORDIERITE) is a precious gemstone with beautiful pleiochromatic effect, meaning it will change color *(typically from violet to blue)* depending on the viewing angle. *The two photos on this page are the same specimen of iolite, taken from opposing angles.* Its name actually comes from the ancient Greek word, "ios", meaning violet. It has been said that Vikings used translucent iolite crystals for navigation. Taking advantage of its natural polarizing effect, an iolite stone could be held up to a cloudy sky and would clearly show brightness radiating from the position where the sun was hidden. Iolite is currently mined in India and Madagascar, yet despite its extreme beauty and durable hardness, it has not become very well-known in the jewelry market. Unfortunately, it is also still uncommon in the collector's market and even small samples are difficult to find and are quite costly.

Healing markets have recently felt increased demand for iolite, perhaps due to the fact that it is purportedly a very high-energy stone. Used by vision seekers and dreamers it can heighten awareness and intuition. It is also a very "social" stone, being useful for maintaining or mending relationships. Hopefully, demand will create more of a market for this beautiful and useful stone.

ASTROLOGICAL: *None,* **CHAKRA:** *Third Eye,* **ELEMENT:** *Water*

IOLITE (IN BLUE LIGHT)

JADE (Jadeite)

FAMILY: Pyroxine Inosilicates

CHEMISTRY: $NaAlSi_2O_6$

HARDNESS: 6.5- 7

STREAK: White

SG: 3.30–3.40

STRUCTURE: Monoclinic Crystal

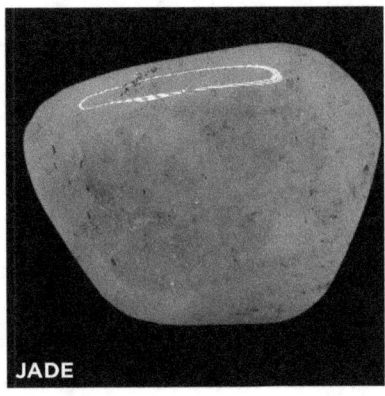
JADE

JADE (JADEITE) & JADE NEPHRITE have historically been almost indistinguishable in name, appearance and in usage, however they are actually different minerals that share many physical properties. Jadeite is a sodium and aluminum silicate pyroxene mineral. Nephrite is a calcium and magnesium silicate amphibole. While jadeite is slightly harder, nephrite is tougher, and both take to carving, shaping or polishing well. Both display a soft luster, a variety of pale colors and a slight pearly translucence. Jadeite and nephrite are considered precious gemstones and have been valued for millennium in the Far East for decoration, jewelry, currency and spirituality. Imperial jade is a vivid, deep green variety of jadeite and is the most highly valued and sought after. Contrary to popular belief, nephrite was traditionally the jade of choice for the Chinese culture until the mid-1700's when abundant deposits of jadeite were discovered in Burma and exported throughout Asia. The finest examples of pure green, translucent jade still come from Burma and a handful of locations in Central America. Recently, the demands of a growing and enthusiastic Chinese market, are making jadeite and jade nephrite rather scarce, and prices have increased dramatically. While new sources for these stones has been found in Japan, Switzerland and California in the U.S.A., they do not match the quality and value of those from traditional sources. Remote areas of British Columbia reportedly have large deposits of jadeite that are relatively untapped.

JADE NEPHRITE

FAMILY: Amphibole Inosilicates

CHEMISTRY: $Ca_2(Mg,Fe)_5Si_8O_{22}(OH)_2$

HARDNESS: 6–6.5

STREAK: White

SG: 2.90–3.02

STRUCTURE: Monoclinic Crystal

GREEN JADE NEPHRITE

JADE (JADEITE) is used for self-protection, business and financial success and dream interpretation. Jadeite has been use for centuries for natural self-healing, treating arthritis and joints, improving bladder function and regulating blood-sugar balance. Recommended by healers for treatment of the eyes, kidneys, joints and for maladies of the lungs.

ASTROLOGICAL: *Aries,* **CHAKRA:** *Heart,* **ELEMENT:** *Water*

JADE NEPHRITE is a stone of wisdom, longevity, fertility, practicality, tranquility and harmony. Physically it is used to improve the body's ability to remove toxins, and to treat problems of the kidneys and adrenal glands. Good for repairing cell damage, it helps cuts and stitches heal with minimal scarring.

ASTROLOGICAL: *Libra,* **CHAKRA:** *Heart,* **ELEMENT:** *Water*

BLUE JADE NEPHRITE

JASPER

SOME OF THE MANY COLORFUL VARIETIES OF JASPER

JASPER

FAMILY: Cryptocrystalline Silicates

CHEMISTRY: SiO_2

HARDNESS: 7

STREAK: White

SG: 2.50–2.91

STRUCTURE: Trigonal Hexagon Crystal

JASPER VARIETIES

JASPER is a common form of chalcedony composed of microscopic silicate crystals *(cryptocrystalline quartz)*. This makes it very similar to its closely related cousin agate. However, whereas agates form predominantly as layered silicate nodules embedded in igneous rock cavities, jasper forms mostly as sedimentary deposits associated with hydrothermal activity, or in silicate-rich aqueous environments such as river beds. The wide range of mottling, spotting or banding in jaspers is due in part to quartz replacement *(pseudomorphism)* of ash, vegetative debris and microbial or bacteriological residues. While the patterns, chemistry and color of jaspers may be quite similar to many agates, it is generally held that jaspers are a more opaque, and a less banded variety of chalcedony than the agates. Jasper stones are abundant, generally inexpensive and readily available, making them easy to collect. We find it very rewarding to seek-out some of the more unusual and uncommon varieties. Jaspers have become a significant and treasured part of our family collection, with over fifty varieties represented.

Jasper stones can be valuable tools in the healer's arsenal as well. Each jasper has unique uses for spiritual and healing purposes. These properties are discussed in more detail in the individual jasper listings on the following pages. One property that all jaspers share, is that they are excellent grounding stones that help us find stability in times of personal turmoil.

Related entries in this book: Agate, Chalcedony, Chrysoprase, Heliotrope, Onyx

BRECCIATED JASPER

DALMATIAN JASPER

BRECCIATED JASPER is a form of red jasper that also contains scattered, black hematite inclusions and frequently, white splotches of colorless quartz or chalcedony. The colorful, broken patches of color are probably due to breaking and refolding of layered jasper due to geologic processes over long periods of time.

ASTROLOGICAL: *Aries,* **CHAKRA:** *Base,* **ELEMENT:** *Earth*

DALMATIAN JASPER is a predominantly white stone with very small flecks of black tourmaline and occasional, orange iron impurities. It is currently found in abundance in Mexico and India, and is very popular with collectors and healers. It is not very costly and is often included in "beginner" rock collections, so it is very well-known, especially with children. Please see the initial listing for jasper for more geological information. All jaspers share general mineralogical and geological data.

Dalmatian jasper encourages an uplifting and positive attitude, bringing good luck and happiness to the bearer. It is used for correcting negative patterns in the daily routine. Reduces feelings of helplessness and brings release from negative issues from the past that may still be influencing your present emotions. Dalmatian jasper instills compassion towards others and patience with those who may need your positive influence. Physically, it is beneficial for recovery from muscle sprains and cartilage damage. Dalmatian jasper promotes peaceful sleep. If you are experiencing nightmares when sleeping, this stone can help reverse the pattern and bring more pleasant and helpful dream states. Children seem to have an attraction to Dalmatian jasper and find it to be a comforting companion.

ASTROLOGICAL: *Aries,* **CHAKRA:** *Base,* **ELEMENT:** *Earth*

DESERT JASPER

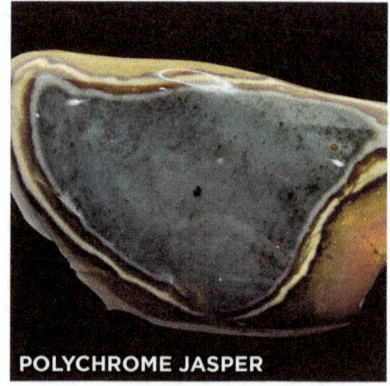

POLYCHROME JASPER

DESERT & POLYCHROME JASPER The terms desert jasper and poly-chrome jasper are often used interchangeably to describe any multi-toned, multi-colored jasper with wide swaths of generally muted hues, and sometimes separated by narrow rings or bands of contrasting color. Most specimens had traditionally come from Australia. Recently, a small but high-quality vein was discovered in Madagascar while miners were searching for the highly prized, orbicular, ocean jasper. Slightly richer in color than Australian desert jasper, and found with attractive naturally domed and curved surfaces, polychrome jasper has become very popular in the retail market. It is predicted that the modest vein deposits of Madagascar polychrome jasper will also soon be depleted, making it even more precious to collectors. The same has happened to the source of ocean jasper. Please see the initial listing for jasper for more geological information. All jaspers share general mineralogical and geological data.

Desert and polychrome jasper encourage ecological sensitivity and awareness, bringing a sense of stability and balance within the physical world. It is a happy and joyous stone, capable of bringing good fortune and a good outlook. As a healing stone it can be used for boosting energy, yet in a calm and efficient manner. It can be used physically to alleviate hay fever and airborne allergy symptoms.

ASTROLOGICAL: *Leo*, **CHAKRA:** *Solar Plexus*, **ELEMENT:** *Fire*

KAMBABA JASPER

KAMBABA JASPER is a beautiful, dark green stone with striking and well-defined black spots and splotches. Darker specimens are often confused with nebula stone because of the orbicular *(round)* inclusions, however nebula stone is an unrelated and much darker-green mineral from Colorado. Kambaba jasper is mined only in Madagascar and is a fossiliferous stone containing silicified* cyanobacteria *(blue-green algae)*, which are responsible for the unusual spotted appearance. Kambaba jasper is estimated to be over 3 billion years old, making it one of the oldest known surface rocks on Earth. Because of its coloring, texture and hardness, it makes a fine semi-precious stone for settings and jewelry. Please see the initial listing for jasper for more geological information. All jaspers share general mineralogical and geological data.

Kamababa jasper offers assistance and protection for astral travelers and those seeking higher enlightenment. It is used to bring mental clarity and focus to all situations. It is useful when studying or reading, as it helps with retention of new information. Physically, it is suggested that Kambaba jasper helps when recovering from long term illness, and improves digestion and boosts the immune system.

ASTROLOGICAL: *Scorpio*, **CHAKRA:** *Heart*, **ELEMENT:** *Earth*

(silicification)—Some minerals or organic materials undergo a process called pseudomorphism, which means that their original chemical structure has been replaced by another mineral. When matter is replaced by quartz or other silicate it is said to have "silicified".

LEOPARDSKIN JASPER

LEOPARDSKIN JASPER is mined primarily in Mexico and a very few other North American locales. It has a beautiful pinkish hue *(orthoclase feldspar)* with multi-colored, orbicular and flecked inclusions, vaguely resembling the skin of a leopard. Technically, this is not a jasper at all, but rather a rhyolite. Jasper would be composed entirely of cryptocrystalline silica quartz; leopardskin is predominantly alkali and plagioclase feldspars which are silicates, but not quartz varieties. Rhyolites such as leopardskin jasper, form in igneous rocks and can have inclusions of several minerals. Slow cooling of the rock causes separation of minerals and provides time enough for nodular clumps or spherical "orbs" to form in the rock. These are present as the characteristic "leopard spots" on this stone. As long as this rhyolite is sold commercially as a jasper, most collectors and healers will continue to refer to it as such, much the same as rainforest jasper, which is also a rhyolite.

Leopardskin jasper helps to strengthen the connection with your spiritual animal totem. It is strongly associated with spiritual discovery and shaman ritual. This stone is useful for self-healing and recovery from all sorts of physical and emotional hardships.

ASTROLOGICAL: *Gemini,* **CHAKRA:** *Base,* **ELEMENT:** *Fire*

Related entries in this book: Rainforest Rhyolite

NOREENA JASPER

NOREENA JASPER is a visually striking stone, with vivid red, orange, yellow, and white geometric shapes. Noreena jasper is formed from compressed, broken and re-solidified mudstones that have silicified* Varying degrees of mineral impurities, such as iron, cause the different colors. It is not a very common variety of jasper, being found perhaps in only one specific, small location in Western Australia. When sold, it is usually in raw, slab state, or as jewelry. Some slabs can be quite attractive and make beautiful artistic displays in the home or collection. Please see the initial listing for jasper for more geological information. All jaspers share general mineralogical and geological data.

Noreena Jasper is said to represent the close relationship between the Earth and human beings. It has been said that the patterns of noreena jasper contain messages from the past and can be read by intuitive and spiritual seekers.

ASTROLOGICAL: *Libra,* **CHAKRA:** *Throat,* **ELEMENT:** *Air*

(silicification)—Some minerals or organic materials undergo a process called pseudomorphism, which means that their original chemical structure has been replaced by another mineral. When matter is replaced by quartz or other silicate it is said to have "silicified".

OCEAN JASPER

OCEAN JASPER is a fine example of an orbicular patterned stone. It was discovered around 1997 in a small, submerged deposit off the coast of Madagascar, which meant that it could only be mined at low tide. In 2006 the last specimen was mined, so no more than what is currently in circulation will ever be available. There are several more common forms of orbicular jasper from various locations throughout the world, however ocean jasper is renowned for its almost infinite variety of colors and its extremely well-defined spherical inclusions. There are several lapidary retailers who still have pieces of this rare stone available, sold in unpolished, slab form. Most of the world's very limited supply of ocean jasper was made into decorative jewelry and some can still be purchased. We are fortunate to have a few small and very precious specimens in our family collection. Please see the initial listing for jasper for more geological information. All jaspers share general mineralogical and geological data

Ocean jasper is an excellent uplifting stone. It helps with positive communications and allows one's true feelings to be more easily expressed in words. It is recommended for use when recovering from physical illness. Working at a cellular level, ocean jasper is a deep-healer of disease and a regenerator of the body's ability to heal itself.

ASTROLOGICAL: *Capricorn,* **CHAKRA:** *Throat,* **ELEMENT:** *Fire*

PICTURE (LANDSCAPE) JASPER

PICTURE (LANDSCAPE) JASPER is also commonly known as landscape jasper. It typically formed from silicified* solutions that dripped into pockets in molten lava, then hardened in banded layers as it became contaminated with different mineral solutions. Because of the plasticity of the igneous base rock, stretching and compression occurred, creating the signature patterns resembling rolling hills, dunes and valleys. Picture jasper is very abundant and easy to find at shops. It is inexpensive and is popular with children, healers and collectors. Because each piece is so unique, it is fun and rewarding to carefully select a picture jasper with a pattern that you love. Even though it does not have bright, bold colors nor striking crystal formations, and is not considered a very precious stone, many of our picture jaspers are favorites in the family collection. Please see the initial listing for jasper for more geological information. All jaspers share general mineralogical and geological data.

Picture jasper encourages ecological awareness, bringing stability and balance within one's surroundings. It is said that every picture jasper contains a message and/or "picture" from the past. Others believe you can see scenes of real Earth landscapes or places we would like to visit. Picture jasper is a stone of balance and harmony and it stimulates creative visualization. It can be used to help heal damage caused by past mistakes in business or in personal relationships. Healers use picture jasper to stimulate the immune system and to treat skin and kidney disorders.

ASTROLOGICAL: *Leo,* **CHAKRA:** *Solar Plexus,* **ELEMENT:** *Earth*

*(silicification)—Some minerals or organic materials undergo a process called pseudomorphism, which means that their original chemical structure has been replaced by another mineral. When matter is replaced by quartz or other silicate it is said to have "silicified".

RED JASPER

YELLOW JASPER

RED JASPER is a stimulating and also an extremely protective stone. It can neutralize radiation and other forms of environmental and electromagnetic pollution. A handy stone to keep near computers and electronics! Red jasper gives strength to confront unfair or abusive situations and grounds the bearer with healthy energy. Red jasper detoxifies the blood and circulatory system, and helps with healthy liver function.

ASTROLOGICAL: *Cancer,* **CHAKRA:** *Base,* **ELEMENT:** *Fire*

YELLOW JASPER is a nurturing stone, that helps to calm the nerves and emotions. It aids with treating disorders of the stomach and digestive system, and is effective in promoting tissue regeneration.

ASTROLOGICAL: *Leo, Chakra: Solar Plexus,* **ELEMENT:** *Earth*

Please see the initial listing for jasper for more geological information. All jaspers share general mineralogical and geological data.

ZEBRA JASPER

ZEBRA JASPER encourages one to think positively and stop considering failure as an option. It is a combination of ethereal white, and grounding black or brown colors, which enable us to balance a perspective between the spiritual and the physical worlds, and to differentiate good from evil. Physically, it is a strong healer that helps with stamina, endurance, and treats bone disorders like osteoporosis. It is suggested for improving teeth, gums, muscle spasms, and heart palpitations.

ASTROLOGICAL: *Leo,* **CHAKRA:** *Base,* **ELEMENT:** *Earth*

Please see the initial listing for jasper for more geological information. All jaspers share general mineralogical and geological data.

JET (Lignite)

FAMILY: Organic Rock

HARDNESS: 2.5–4

STREAK: Brown

SG: 1.30–1.34

JET

JET (LIGNITE) is not technically a rock or mineral, but is an organic mineraloid called lignite, which is a low-grade relative of coal. It is the fossilized
product of ancient tree wood from millions of years ago. It is found in hard
and soft forms. Both are the result of compression of carbon-rich substances,
but hard jet formed in a salt water environment, and soft jet in fresh water. The
major source of jet since Roman times has been the cliffs of Whitby on the west
coast of England. It was historically popular for wearing as somber, mourning
jewelry. It is hard enough to take a nice polish and it can be carved and formed
easily. Ornamental jewelry made from jet has been discovered in Spain that
dates back to 17,000 B.C., and in Germany dating back to 10,000 B.C. The term
"jet black" derives from the appearance of this substance as it is truly a deep,
solid and true black color. Polished specimens of jet are extremely reflective
with a high degree of shine.

Jet is used for spiritual purification and protection. It clears all chakras and
replaces bad energy with positive life-force. If kept in a bowl or container with
your healing stones, it will keep them cleansed and charged. This is particularly
useful for crystals that should not be exposed to sunlight or rinsed in water.* For
physical healing purposes, jet is used to reduce fever and the aches and discomfort that accompany colds and flu. Jet stabilizes the body's equilibrium and
stops dizziness.

ASTROLOGICAL: *Capricorn,* **CHAKRA:** *Base,* **ELEMENT:** *Air*

Related entries in this book: Amber, Copal

*Please refer to the section in this book for "Care and Cleansing of Crystals"

K2 STONE

FAMILY: mixed mineral

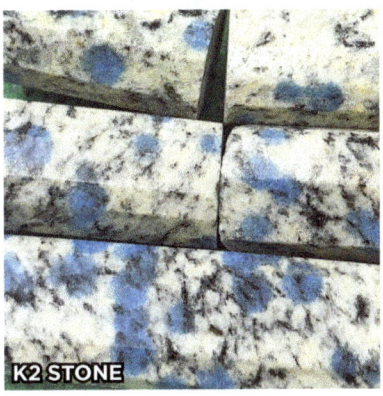

K2 STONE is a recently created name for marketing of azurite in granite matrix. It is sometimes incorrectly referred to as K2 jasper. It is in no way related to jasper. The azurite is embedded in the granite in spherical form, usually between 1/8"–1/2" diameter, so when the stone is cut or slabbed, the spheres appear as blue circular shapes on the cut surface. Typically the granite is mainly white feldspar with some content of sparkling quartz, with very small flecks of biotite and muscovite micas. It's quite a striking display of bright blue circles in lightly fine-textured and lightly colored granite. K2 stone is mined within sight of, and named for, the second highest mountain in the world, near the China/Pakistan border.

Being rather recently introduced to the world, crystal healers are still finding uses for this unique stone. I have seen evidence of it being used for enhanced intuition, spiritual connection, Akashic learning, and inter-dimensional experience. Others use K2 for seeking connections to the archangels.

ASTROLOGICAL: *Gemini,* **CHAKRA:** *Crown,* **ELEMENT:** *Fire*

Related entry in this book: Azurite

KYANITE

FAMILY: Nesosilicates

CHEMISTRY: Al_2SiO_5

HARDNESS: 4.5 parallel to crystal, 6.5 perpendicular to crystal

STREAK: White

SG: 3.53–3.88

STRUCTURE: Triclinic Crystal

BLUE KYANITE IN MICA

KYANITE is an aluminum-rich, silicate mineral that occurs in many vivid colors, most commonly blue, black and green but sometimes also shades of red and orange. It is found in sedimentary and metamorphic rocks but most frequently in igneous pegmatite intrusions, which is true of many other large, precious gem crystals. It is highly anisotropic, meaning that its hardness varies with orientation to its crystal grain *(harder across-grain than with the grain)*. It can be found around the world in locations such as Brazil, Norway, Kenya and Myanmar and also in New Hampshire and California, USA. Natural, fan-shaped growths of bladed crystals make an attractive display piece for the collection.

Kyanite is a solid healing tool for many purposes. It is purported to help higher brain function, and lower blood pressure. It is used by healers as a natural pain reliever, especially for arthritis and rheumatism. Kyanite brings a sense of calm, which encourages inherent psychic abilities to surface, and develops communication on all levels. It helps drive away anger and stress and promotes logical thought. Kyanite *(especially the blue variety)* brings lucid dreams and retention of those dreams. It helps calm the nerves of busy students and new drivers, making it a useful healing crystal for the whole family.

ASTROLOGICAL: *Aries,* **CHAKRA:** *Base,* **ELEMENT:** *Earth*

BLACK KYANITE

GREEN KYANITE

ORANGE KYANITE

LABRADORITE

FAMILY: Tectosilicates

CHEMISTRY: $(Ca,Na)(Al,Si)_4O_8$

HARDNESS: 6–6.5

STREAK: White

SG: 2.68–2.72

STRUCTURE: Triclinic Crystal

LABRADORITE

LABRADORITE is member of the plagioclase feldspar family. Its higher calcium content differentiates it from other aluminum silicate feldspars such as orthoclase, which are potassium rich. It tends to form in mafic *(silicate-poor)* igneous rock formations and is commonly found in association with olivine, pyroxenes, and amphibole minerals. The main source for labradorite is Labrador, Canada, specifically Paul Island, and small deposits can be found in Norway and Finland where is usually referred to as specularite. It is desired for its famous multi-hued shimmer and shiller, referred to as "labradorescence". This gives a good specimen the appearance of changing hue from dull gray to striking blues, greens or gold when viewed from different angles. It is a very hard mineral, yet lends itself well to polishing or even slicing, which is a popular method of displaying its beautiful color-shifts. I have never met anyone who is not in awe at the beauty of a good specimen of labradorite. Wonderful stuff!

Labradorite helps to clarify which tasks and issues which need to be addressed, and brings better ability for successfully completing them. It balances the aura and the chakras, raises consciousness, and stabilizes spiritual energy, making for productive and deep meditative states. It strengthens intuition and is helpful for attuning psychic abilities. It can reveal the truth behind illusions, releasing fear and insecurity. Labradorite boosts enthusiasm for learning and helps with acceptance of new ideas. Great for students! Labradorite treats the eyes and brain, stimulates mental acuity, and relieves anxiety and stress. It regulates metabolism, balances hormones and relieves tension from moon-cycles. It treats common colds, gout, and rheumatism, lowers blood-pressure, and aids in digestion.

ASTROLOGICAL: *Leo,* **CHAKRA:** *Base,* **ELEMENT:** *Air*

Related entries in this book: Feldspar, Moonstone, Sunstone

LAPIS LAZULI (Lazurite)

FAMILY: Phosphates

CHEMISTRY: $(Na,Ca)_8[(S,Cl,SO_4,OH)_2$

HARDNESS: 5.5–6

STREAK: White

SG: 2.40–2.91

STRUCTURE: Isometric Crystal

LAPIS LAZULI

LAPIS LAZULI (LAZURITE) has been mined continuously in Afghanistan for over 6000 years. Traditionally, the pigment aquamarine was made from ground and powdered lazurite. Other blue minerals such as azurite and sodalite are often confused with lazurite. To clarify, lazurite is a mineral; lapis is a rock and includes minerals such as lazurite, sodalite, pyrite, hayune and others. Being fairly rare, lapis is highly prized as ornamentation and jewelry, and is valued accordingly. It rarely forms crystals of any great size, but as a rock it is easily worked for decorative purposes. Lapis is easy to find on the collector's market and in jewelry stores. It is fairly expensive and better quality stones fetch a higher price. Usually deep, rich blue is preferable for jewelry purposes.

Lazurite guards against psychic attacks. It quickly releases stress, bringing deep and restful peace. It brings harmony and deeper self-knowledge, revealing true awareness of one's own psyche. Lazurite allows free self-expression and reveals inner-truths to the user. Used to boost the immune system, purify blood and lower blood pressure, it also soothes and relieves areas of inflammation and irritation. Lazurite helps alleviate insomnia and depression and it is of benefit to the respiratory tract, the nervous system and the throat.

ASTROLOGICAL: *Taurus,* **CHAKRA:** *Throat,* **ELEMENT:** *Earth*

LARIMAR

FAMILY: Inosilicate

CHEMISTRY: $NaCa_2Si_3O_8(OH)$

HARDNESS: 5

STREAK: White

SG: 2.74

STRUCTURE: Triclinic

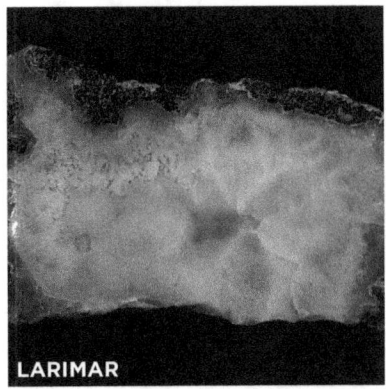

LARIMAR

LARIMAR is a very beautifully colored blue mineral, and a semi-precious gemstone variety of the mineral percolate. It is mined exclusively in the Dominican Republic. It was reportedly first found there around 1916 but has only been commercially mined since the mid 1970's. Being as it comes exclusively from a single locale, and is a relatively recent find, larimar is rather costly and comes to the market intermittently. Larimar is too soft to be used for some jewelry purposes but is fashioned into pendants and rings, usually surrounded by Sterling silver or other metals to protect it from chipping. It is opaque and has little reflectivity, but it's soft blue, to bluish and green turquoise shades make it a very attractive semi-precious gemstone.

For crystal healing or spiritual endeavors larimar is used as a "speakers stone" channeling energy through the throat chakra. Thus, it enhances one's ability to be heard, to counter negative speech from others, and maintain personal power. Similar to its placid blue coloration, larimar is used for the physical effects of calming and cleansing the spirit. Reflexologists and acupuncturists sometimes like to have the stone handy for relaxing their subjects while performing their work.

ASTROLOGICAL: *Pisces,* **CHAKRA:** *Throat,* **ELEMENT:** *Water*

LARVIKITE

FAMILY: Igneous Rock

LARVIKITE

LARVIKITE is an igneous rock and a variety of the mixed-mineral monzonite. Small embedded crystals of feldspar give larvikite a stunning sparkle and occasional labrador essence *(flashes of blue)*. It is not a hard stone but takes a polish very well which enhances its color and shine. The tiny sparkles resemble copper or bronze but these metals are not a part of its structure. Titanium flakes may be seen in some specimens. Larvikite is found as part of a large igneous batholith structure in the region of Larvik, Norway, from which it obviously gets its name.

Larvikite is not common at all as a small specimen stone or as a healing crystal, however surprisingly it is mined and cut into huge slabs which go mainly to industries that produce countertops, sinks and wall tiles. Large slabs are also used on building facades, which is a stunning way to display such a "flashy" stone! More uses such as statuary and adornment are being created as more of the material is being mined.

Crystal healing has found larvikite to be useful for protection and establishing personal boundaries. It is considered to be a grounding stone for the spirit and makes a connection between the spiritual self and the physical Earth.

ASTROLOGICAL: *Virgo*, **CHAKRA:** *Heart*, **ELEMENT:** *Water*

LEMURIAN SEED CRYSTAL

FAMILY: Quartz Silicates

CHEMISTRY: SiO_2

HARDNESS: 7

STREAK: White

SG: 2.58–2.59

STRUCTURE: Trigonal Hexagon Crystal

LEMURIAN SEED CRYSTAL

LEMURIAN SEED CRYSTAL, as legend has it, 12 million years ago there existed an advanced civilization somewhere between the American continent and Australia. Called the Lemurians, their society was peaceful, intelligent and highly developed spiritually. During the last days of Lemuria, it was decided to plant "seed crystals" that were programmed to transmit a message of love and unity, and to pass-on to the future, their accumulated knowledge of hundreds of generations. Having seeded the crystals, some Lemurians, it is believed, left this planet for other star systems. Others went into inner earth *(while maintaining telepathic connection with those in other parts of the galaxy)*, where they continue to care for the planet and the seed crystals that are now surfacing.

Lemurian crystals are said to have been first found in Brazil in 1999 lying loose in a bed of sand, unattached to base rock, which would be a normal growth habit for quartz crystals. Odd as this is, the physical appearance of these crystals is even more interesting. Many are not entirely transparent and shiny, but have almost a frosted appearance. Some have a slight pink glow. Striations reminiscent of etchings or hieroglyphs appear on some crystals and small embedded triangles appear in others. Etched specimens are said to hold ancient knowledge that will in time, be revealed to the current bearer. Lemurian crystals with triangular markings on the surface are said to be healing stones of great power. Pink hue is indicative of extensive use in the past. Lemurian crystals have become quite popular for meditation and spiritual practices and are available from a wide range of sources.

ASTROLOGICAL: *Virgo,* **CHAKRA:** *Third Eye,* **ELEMENT:** *Air*

NOTE: *All conjecture about Lemurian crystals is considered by this author to be anecdotal in nature and is included in this guide purely to illuminate an interesting point of view, and because the crystals have some uses to spiritualists and crystal healers. Like any good quartz prism, Lemurian crystals can be quite beautiful, but this author would caution the buyer against paying more money for a Lemurian crystal than any other good quartz crystal.*

LEPIDOLITE

FAMILY: Phyllosilicates

CHEMISTRY: $K(Li,Al_3)(Al,Si,Rb)_4O_{10}(F,OH)_2$

HARDNESS: 2.5-3

STREAK: White

SG: 2.80-2.90

STRUCTURE: Monoclinic Crystal

LEPIDOLITE

LEPIDOLITE typically occurs in granite intrusions or in quartz deposits. It is a member of the mica family but displays somewhat less tendency to cleave into flakes or sheets than other varieties. Its gentle purple or pink appearance is due to its high lithium content, which also makes it a valuable commercial source of lithium. It also occasionally contains the rare metals cesium and rubidium. Lepidolite specimens are available to collectors in a wide variety of hue and form. It polishes fairly well for such a soft stone, and it is also attractive and colorful in its raw state. Less commonly it can be found as a peculiar botryoidal *(globular)* mass, which this photograph from our family collection displays beautifully.

Lepidolite is a good stone for meditation, as it facilitates heightened awareness and balance of spirit. It promotes kindness and understanding of others. Physically, it boosts the immune system and protects against allergic reactions. It is a stone of support when recovering from traumas and for those trying to break harmful habits.

ASTROLOGICAL: *Libra,* **CHAKRA:** *Third Eye,* **ELEMENT:** *Earth*

Related entries in this book: Fuchsite, Muscovite, Star Mica

LIBYAN GLASS

FAMILY: Silicates

CHEMISTRY: SiO_2

HARDNESS: 5

STREAK: None

SG: 2.6

STRUCTURE: Amorphous Glass

LIBYAN GLASS

LIBYAN GLASS is a beautifully translucent tektite, typically and uniquely tinted with hues of gold and yellow. Some specimens are shaped as discreetly shaped "bombs" indicating a ballistic trajectory of molten material cooling in mid-air, after an impactor hits the ground, dispersing molten silicates and other minerals through the air at high-velocity. Other samples look much like broken beach-glass, with slightly frosty surfaces and irregular shapes. This would indicate shattering of larger tektites, with subsequent surface-alteration, perhaps due to the sandy nature of the locations Libyan Glass tektites are found.

There does not seem to be geologic evidence for the exact impact site that would be the origin of the Libyan tektites, but inclusions of trace amounts several exotic minerals, confirm an extraterrestrial impactor was involved in their formation.

Libyan tektites have been collected from the desert sands of Western Egypt and Libya for at least 3000 years, as evidenced by jewelry and carvings of the material used in the time of King Tut.

ASTROLOGICAL: *Aries,* **CHAKRA:** *Base,* **ELEMENT:** *Air and Fire*

Related entries in this book: Moldavite, Tektites

MAGNESITE

FAMILY: Carbonate

CHEMISTRY: $MgCO_3$

HARDNESS: 3.5–4.5

STREAK: White

SG: 2.98

STRUCTURE: Trigonal

MAGNESITE

MAGNESITE is a common carbonate material that has formed mostly near the contact zones between crustal and mantle materials where the occurrence of magnesium is higher than at the surface of the Earth. Magnesite is found in a very wide range of locations on Earth and has even been detected on Mars. Because it is a rather soft mineral, magnesite has been used extensively throughout human history for decorative purposes; carvings and beads being most common. Perhaps sadly, much of the magnesite used today for commercial and decorative purposes is to mimic another more valuable and desirable mineral, howlite. In raw form, magnesite can look almost identical to howlite, and also because of its porosity, takes artificial dye very easily, leading to many brightly colored specimens, chiefly blue, posing and being sold as howlite.

ASTROLOGICAL: *Aries,* **CHAKRA:** *Crown,* **ELEMENT:** *Earth*

MALACHITE

FAMILY: Carbonates

CHEMISTRY: $Cu_2(CO_3)(OH)_2$

HARDNESS: 3.5–4

STREAK: Light Green

SG: 3.70–4.01

STRUCTURE: Monoclinic Crystal

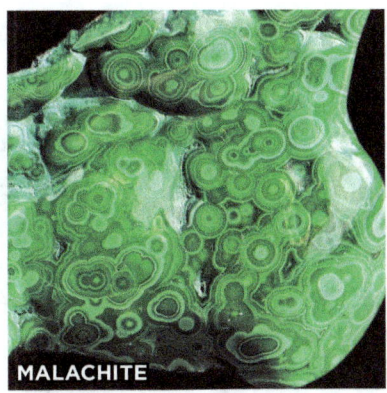

MALACHITE

MALACHITE is a beautiful, semi-precious gemstone that has been used for ornament and architectural detail for centuries. Its characteristic swirled patterns are caused naturally when malachite forms as a botryoidal deposit or layered, lumpy mass. Repeated mineral deposits and oxidation cycles cause the variations in color and make it an ideal stone for cutting or polishing. Malachite often results from weathering of copper ores and is frequently found in association with azurite in copper ore deposits. Except for the green color, the physical and chemical properties of malachite are almost identical to those of azurite, which is brilliant blue. Furthermore, in time and with exposure to air, azurite will eventually oxidize further and morph into malachite. Large quantities of malachite have been mined in Russia, Gambia, Congo, Israel and the United States. Typically available on the market are masses of malachite which are surface polished to enhance color and pattern *(as shown above)*. Malachite specimens are the centerpiece of many collections, including our own. It really is incredibly beautiful!

Malachite aids concentration and releases the bearer from the after-effects of negative energies. It encourages one to take only proper and healthy risks and make sound decisions. Malachite helps release stress and anxiety that comes from overworking and over-stimulation of the senses. Healers use malachite to strengthen the immune and nervous systems, to lower blood pressure, to treat asthma, arthritis, epilepsy, bone fractures, swollen joints, and travel sickness. Malachite helps focus healing energy to target tumors and cancerous cells. It is an excellent stone for concentration and an aid when studying or doing homework.

ASTROLOGICAL: *Libra*, **CHAKRA:** *Heart*, **ELEMENT:** *Earth*

Related entries in this book: Azurite, Chrysocolla

MANUFACTURED GLASSES

Manufactured glasses are obviously not geologic entities nor rocks, but they do have several applications for decorative use, and are widely used in the endeavors of crystal healing. Thus, some of the more popular manufactured glasses are included here.

SEE PAGE #S BELOW FOR INDIVIDUAL MANUFACTURED GLASS

MARCASITE

FAMILY: Sulfide

CHEMISTRY: FeS_2

HARDNESS: 6

STREAK: Black

SG: 4.9

STRUCTURE: Orthorhombic

MARCASITE

MARCASITE is the lesser known sulfide of iron in both mineral collections and crystal healing arsenals, it's sister mineral pyrite being far more common. Ironically, when pyrite has been used for jewelry purposes it is traditionally called marcasite. Marcasite and pyrite together are the most important and abundant commercial ores of iron, and are found abundantly around the world. Small specimens of marcasite do not take a polish quite as well as pyrite which may the reason it is not collected as commonly. Still, beautifully formed marcasite crystals can be found in association with fluorite, calcite, dolomite, and galena, and make excellent cabinet specimens that rival far more exotic and expensive crystals.

Healers traditionally used marcasite for seeking self-awareness, deep meditation and wisdom. Ancient Romans wore talismans or carved figures made of marcasite to represent courage and strength. Modern uses are for focus and fighting fatigue, so it is commonly suggested for students as a study aid.

ASTROLOGICAL: *Scorpio,* **CHAKRA:** *Third Eye,* **ELEMENT:** *Earth*

Related entries in this book: Hematite, Pyrite

METEORITES

CHEMISTRY: Typically (Fe,Ni)2Si

HARDNESS: 5.0–6.5

STREAK: None

SG: Varies

METEORITES

METEORITES present the collector with perhaps the most unique mineral specimens available on the planet! Most samples will be small, as meteors tend to break-up in the atmosphere before reaching ground, and also because more massive examples usually end-up in museums. A meteorite can have variable composition and are divided into two general categories; carbonaceous chondrites and nickel-iron.

Chondrites are the more "primitive" of the two and show traces carbon and silicate, and some have a small degree of amino acids *(life building chemicals)*. Chondrites formed very early in the origin of the solar system and thus, were not subject to high-temperature alteration that went into forming planetary bodies.

Nickel-iron meteorites are in some ways, the opposite. They are the remnants of heat-altered material and have a higher degree of metals. Any traces of carbon or amino-acids, etc. would have been "cooked" out of these bodies as large bodies formed from the pre-solar nebula.

All meteorites collected will show melted, irregular surfaces from their travel though our atmosphere at high speed. Most meteorites are dust size, larger ones tend to explode in the atmosphere and rain down a variety of sizes.

For crystal healing, meteorites have a very wide range of uses. Most commonly they are suggested for bringing order out of a chaotic situation, calming nerves, soothing the soul. Others tap into the higher energies of the celestial object and use meteorites for vision-quest, clairvoyance, and psychic work. Meteorites have been recommended for energizing all chakras, and they can be of relevance to people of all astrological signs.

MICAS

Micas are a widespread and important family of minerals, found in many rocks and mixed-minerals, and have many uses both practical and ornamental. Micas such as biotite are a constituent of granite which is a major component of Earth's crust.

But for the purpose of this book, we focus on the mica varieties that themselves make-up discrete and separate minerals. The variety of shapes, colors, and other physical features that these micas display are obvious when you compare the listings in this entry.

SEE PAGE #S BELOW FOR INDIVIDUAL MICAS

MOLDAVITE

FAMILY: Silicates

CHEMISTRY: $SiO_2(+Al_2O_3)$

HARDNESS: 5

STREAK: None

SG: 2.6

STRUCTURE: Amorphous Glass

MOLDAVITE

MOLDAVITE is a deep-green tektite, found mostly as small fragments of irregular shape and varied texture. The main source for finding this material is the Moldau region of Czech Republic, and has become wildly popular with collectors and jewelers the last two decades. There is geologic evidence that the impact that created these tektites was the Nördlinger Ries crater in Germany, approximately 15 million years ago.

Sadly, there is a considerable amount of fake or imitation moldavite circulating the mineral and jewelry markets. Moldavite is expensive and precious, but green bottle-glass and green plastics, have been altered to closely resemble the naturally formed tektite. When considering adding moldavite to your collection, please buy only from a reliable dealer or ask an expert. Authentic moldavite tektites are a wonderful addition to any collection!

ASTROLOGICAL: *Aries,* **CHAKRA:** *Base,* **ELEMENT:** *Air and Fire*

Related entries in this book: Libyan Glass, Tektites

MOOKAITE

FAMILY: Tectosilicates

CHEMISTRY: SiO_2

HARDNESS: 7

STREAK: White

SG: 2.64–2.69

STRUCTURE: Trigonal Hexagon Crystal

MOOKAITE

MOOKAITE has been classified many ways and is most commonly referred to as a type of jasper. However, its water content varies, and the fact that it may have derived from biological matter, means it could possibly be considered a chert or opal mineral. Mookaite is found in only one deposit; the dry Mooka River bed in Australia. It was formed primarily from layers of ancient bacterial deposits in clay-mud that have been replaced by silicate minerals from river water. Impurities of other minerals turned the stone to various rich shades of red, purple and yellow. Mookaite is very hard and takes a nice polish. It is mined and distributed annually by the owner of the Australian river bed, and while it not extremely abundant, it is readily available commercially. There are several common jaspers and agates that are often mislabeled and sold as mookaite, probably due their similarity in color. Mookaite comes in shades of magenta, red, white, cream and yellow. Often, all of its colors can be found in one small sample. The yellow variety of mookaite is one of the few truly yellow stones available. Mookaite has become a family favorite and we have specimens of all sizes and shapes, both polished and raw.

Mookaite blocks negative energy that can lower the energy level of the spirit. It brings an embracing love for new situations and promotes spiritual growth. Mookaite is a boost to the immune system and is beneficial especially during the cold and flu season, maintaining the body's general health and wellness. A very "earthy" stone, mookaite can be used for communicating with nature, and it makes for a great companion in the garden. We have placed several mookaites in our garden, near plants that have trouble weathering heat and drought.

ASTROLOGICAL: *Leo,* **CHAKRA:** *Heart,* **ELEMENT:** *Water*

Related entry in this book: Jasper

MOONSTONE (Adularia)

FAMILY: Tectosilicates

CHEMISTRY: $(Na,K)AlSi_3O_8$

HARDNESS: 5.5-6

STREAK: White

SG: 2.54-2.59

STRUCTURE: Monoclinic Crystal

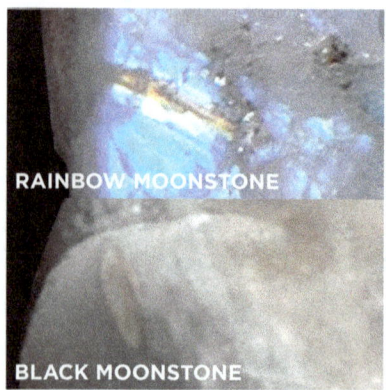

RAINBOW MOONSTONE

BLACK MOONSTONE

MOONSTONE (ADULARIA) is an oligoclase feldspar mineral, part of the sodium and calcium-rich plagioclase feldspar family and is also known mineralogically, as adularia. Its name comes from the opalescent glow caused by successive layers of semi-translucent feldspar. It is a fairly hard stone due to its high silica and aluminum content. Varying degrees of aluminum in relation to feldspar create different hues ranging from pearly white to soft gray to light pink. It makes an attractive tumbled and polished stone and has been used for adornment and jewelry for centuries. Moonstone can be found in Australia, Burma, Norway, Brazil and many other locations making it fairly common and easy for collectors to find. Moonstone is the official, state mineral of Florida even though it does not occur there naturally, but is an honorary title in reference to NASA's space program.

Moonstone encourages deep thought and reflection, helping one to make sound and logical decisions. It also heightens awareness and brings clarity to vision seekers. Moonstone is an aid to creative, artistic and musical creation due to its strong ability to encourage free expression. Moonstone is also known to help with digestion and protect the digestive organs, most specifically the liver and kidneys. Moonstone is suggested as an aid for absorption of vitamins and minerals. Moonstone has traditionally been used to balance hormonal swings and moon cycles.

ASTROLOGICAL: *Cancer,* **CHAKRA:** *Third Eye,* **ELEMENT:** *Water*

Related entries in this book: Labradorite, Orthoclase Feldspar, Sunstone

PEACH MOONSTONE

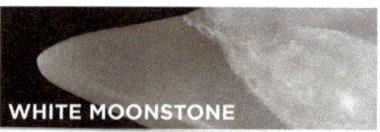

WHITE MOONSTONE

MORGANITE

FAMILY: Cyclosilicates

CHEMISTRY: $Be_3Al_2(SiO_3)_6$

HARDNESS: 7.5–8.0

STREAK: White

SG: 2.74–2.76

STRUCTURE: Hexagonal Crystal

MORGANITE

MORGANITE is a light-pink beryl, from the same crystal family as the more popular emerald and aquamarine. While a perfectly clear, inclusion-free emerald is practically non-existent, morganite, like heliodor, can be nearly flawless and is therefore quite precious. Morganite can be found in association with other colors of beryl in many granite pegmatite formations in several places throughout the world. It was first discovered in California in 1910 by George Kunz, a notable mineralogist *(the mineral kunzite is named for him)* who at the time was an employee of J.P. Morgan, the wealthy financier. Kunz named the newly discovered beryl after Mr. Morgan, due to Morgan's importance as both an avid gem collector and as a financier of several large-scale mining operations. Morganite is one of the least used beryls for jewelry purposes, and is also somewhat difficult to obtain as a specimen even in lower quality form. We were able to obtain a small, raw specimen of good color and fair clarity from a reputable, online dealer of minerals.

ASTROLOGICAL: *Libra,* **CHAKRA:** *Heart,* **ELEMENT:** *Water*

Related entries in this book: Aquamarine, Emerald, Goshenite, Heliodor

MOQUI MARBLE & SHAMAN STONE

FAMILY: Mixed Mineral Concretion

CHEMISTRY: SiO_3 inside Fe

HARDNESS: 4.5

STREAK: Bright to Dark Red

SG: Varies

STRUCTURE: Mixed Crystal Structure

MOQUI MARBLES

MOQUI MARBLE The word "Moqui" comes from Hopi language and means "the dead". It is said that ancestors would come to Earth and play with these "marbles" at nighttime to be near the living. At sunrise, the spirits would return to their realm, but left the marbles behind as a reminder. Moqui marbles actually come from the famous Navajo Sandstone Formation in Utah. Spread across the desert-floor, they formed as round or elliptical concretions of compacted sandstone, with a shell of hematite. The process is not entirely clear, but they probably formed in cavities within the walls of sandstone that were saturated by iron-bearing water which encased the sandstone and created these spherical concretions. Due to weathering, they became exposed and eventually fell out of the rock wall and settled in alluvial beds. The area where moqui marbles are collected has recently become protected, so no more collecting will be allowed. Some are still available, but most of the best specimens are gone.

Moqui marbles are believed to be among the most energetic stones on earth. They are a strong aid while meditating or achieving higher states of mind. They assist with forming bonds with past and the future and with spirit guides from all levels of existence. At the same time, moqui marbles are, relaxing and stabilizing. Moqui marbles have either male or female energy. Male stones are typically ridged or rough and may be slightly flattened spheres and give energy to the lower half of the body. Female stones are typically smooth and round and bring energy to the upper body. Moqui marbles are often used in pairs to balance the chakras in the entire body, and to balance the yin/yang of the spirit. Many people describe a pair of moqui marbles tugging towards each other when held in opposite hands, yet they have no natural magnetic properties. We have experienced this ourselves and with friends and family members, and it is truly remarkable!

ASTROLOGICAL: *Libra,* **CHAKRA:** *All,* **ELEMENT:** *Earth*

Related entries in this book: Boji Stone™ , Prophecy Stone, Septarian

MUSCOVITE (Mica)

FAMILY: Phyllosilicates

CHEMISTRY: $KAl_2(AlSi_3O_{10})(F,OH)_2$ SiO_2

HARDNESS: 2 Perpendicular & 4 Parallel to Crystal Grain

STREAK: White

SG: 2.75–2.73

STRUCTURE: Monoclinic Crystal

MUSCOVITE

MUSCOVITE (MICA) is a class of phyllosilicate minerals that forms under extreme pressure. Crystal alignment and plate-like structure indicates past lateral movement and is a good indicator mineral for geological conditions in the past. Muscovite masses are generally soft, flexible and can flake into thin sheets. Other familiar members of the mica family are biotite, vermiculite, fuchsite and lepidolite. Muscovite is the most common type of mica and can be found in numerous locations throughout the world. India is the leading producer of muscovite mica followed by Brazil. Its name is Russian and derives from its former use in Moscow and surrounding regions for window glass *(Muscovite Glass)*. It can be found generally in granite intrusions, pegmatites or schists. It is most commonly found in shades of pale green and gray and occasionally pink. It has been useful in the past as a substitute for glass in high-heat conditions such as oven windows. It is still of great commercial value as an insulator for both heat and electricity. It separates easily into extremely thin sheets which are highly transparent. Muscovite is flexible, bending easily and returning to shape, and is stronger perpendicular to its grain structure than it is parallel. Sheets that are thick enough to remain opaque to light can have a high level of reflectivity and a mirror-like silver sheen.

Muscovite has many spiritual uses but because of its fragile nature is rarely used. It gives encouragement to heed spiritual messages and spirit-guides and to address the higher-self. It brings all aspects of the mind into tune, making it useful for tasks which require full attention, such as studying or testing. Muscovite can help ease physical pain and muscle tension, as it increases agility and flexibility.

ASTROLOGICAL: *Cancer,* **CHAKRA:** *Sacral,* **ELEMENT:** *Air*

Related entries in this book: Fuchsite, Lepidolite, Star Mica

NUUMMITE

FAMILY: Amphibole Inosilicates

CHEMISTRY: $(Mg_2)(Mg_5)Si_8O_{22}(OH)_2$

HARDNESS: 5.5–6

STREAK: White

SG: 2.92–3.03

STRUCTURE: Orthorhombic Crystal

NUUMMITE

NUUMMITE is typically dark-black and opaque with tiny inclusions of metallic pyrite and/or chalcopyrite flakes that give it a deep sparkle and shine. It is composed primarily of two amphibole minerals, gedrite and anthophyllite, which form in layers, giving nuummite a degree of iridescence. Some better examples show slight pleochromism, appearing black from one angle and shimmering bronze from another. It is a fairly rare stone, mined mostly in Greenland during the warm months and is named for the Nuuk region where it is found. Nuummite is a very old metamorphic rock *(nearly 3 billion years old)*, making it much older than most surface minerals on the Earth's crust. Its primary use is as a decorative stone and for jewelry and has become very popular with spiritualists and crystal healers, as polished palm-stones have recently become more readily available.

Nuummite has long been known as "the sorcerer's stone" and is the revealer of hidden truths. It is an aid to clarity of vision and induces communication in higher states. It is of great use for deep meditation. It is a valuable and durable protection stone that can shield the bearer from negative energy directed at both the body and the emotions. Nuummite balances intellect and intuition, which together, are a potent force when dealing in business. These traits also make it useful for revealing secrets and solving riddles. Nuummite brings good luck and can help find lost items.

ASTROLOGICAL: *Sagittarius,* **CHAKRA:** *Third Eye,* **ELEMENT:** *Air*

OBSIDIAN

FAMILY: Mineraloids

CHEMISTRY: SiO_2 + MgO, Fe3O$_4$

HARDNESS: 5.5

STREAK: White

SG: 2.40–2.60

STRUCTURE: Amorphous Rock

OBSIDIAN is volcanic glass that forms quickly from molten lava. Rapid cooling does not allow time for the silicate molecules to align themselves into a true crystalline form. Thus, obsidian is technically considered a glass, not a rock. For most mineralogical classification purposes, obsidian is placed in a general category called mineraloids. Glasses *(obsidian, Apache tears, fulgarite)*, organics *(amber, copal, jet)* and concretions *(moqui marble, boji stone, septarian)* are all members of the mineraloid family. There are numerous colors and forms of obsidian; some opaque, some transparent. All obsidians share similar chemistry, with silica being the main constituent. Impurities of metals or oxides are responsible for variations in color.

BLACK OBSIDIAN

MAHOGANY OBSIDIAN

BLACK OBSIDIAN helps to discern truth among confusion. It is strongly protective and defends against negativity and psychic attacks. It absorbs and eliminates negative environmental energy. Black obsidian is suggested as an aid to digestion and it detoxifies the entire body. It can reduce pain from arthritis, joints and moon-cycle cramps. Black obsidian brings a sense of warmth to the body.

ASTROLOGICAL: *Scorpio,* **CHAKRA:** *Third Eye,* **ELEMENT:** *Earth*

MAHOGANY OBSIDIAN can be used as a guide towards making proper decisions and healthy choices. It has its own positive energy which transfers to the bearer. Mahogany obsidian and red obsidian are suggested for health of teeth and gums, and is good for overall health of the mouth, tongue and throat.

ASTROLOGICAL: *Scorpio,* **CHAKRA:** *Heart,* **ELEMENT:** *Earth*

SILVER SHEEN OBSIDIAN

SNOWFLAKE OBSIDIAN

SILVER SHEEN OBSIDIAN is a very protective stone that forms a shield against negativity and absorbs negative psychic energy. It encourages patience and perseverance. It can be used to focus meditation and to help maintain a conscious connection to the physical body during astral projection. Silver sheen obsidian is recommended for improving eyesight and alleviating eye strain.

ASTROLOGICAL: *Pisces*, **CHAKRA:** *Third Eye*, **ELEMENT:** *Earth*

SNOWFLAKE OBSIDIAN improves blood circulation and encourages the re-growth of cells and skin. It can be used to focus blurry vision and elevate moods. Healers use snowflake obsidian as a natural pain reliever and energizer. It is especially considered for easing the discomfort and pain of migraine headaches.

ASTROLOGICAL: *Scorpio*, **CHAKRA:** *Heart*, **ELEMENT:** *Earth*

OBSIDIAN GLASS

FAMILY: Manufactured Glass

CHEMISTRY: SiO_2

HARDNESS: 5

STREAK: None

SG: 2.6

STRUCTURE: Amorphous Glass

OBSIDIAN GLASS

OBSIDIAN glass is quite simply, exactly what it's name implies—it is a non natural substance. As such, it would hardly merit inclusion in a mineral or geologic guidebook, however the uses for non-natural glass objects are extensive in the crystal healing world, and they do also make attractive pieces in rock collections. Please remain aware, that these glasses are never formed naturally.

The obsidian glasses sold at rock shops are a high-quality silicate glass, silicate also being the main constituent of natural glasses such as tektites *(impactites)* and volcanic obsidians. Chemical colorants are used to create rich, deep colors; green, blue, and yellow, being most common.

ASTROLOGICAL: *Scorpio,* **CHAKRA:** *Heart,* **ELEMENT:** *Earth*

Related entries in this book: Goldstone, Opalite

ONYX

FAMILY: Cryptocrystalline Silicate

CHEMISTRY: SiO_2

HARDNESS: 6.5–7.0

STREAK: White

SG: 2.58–2.64

STRUCTURE: Hexagon Crystal

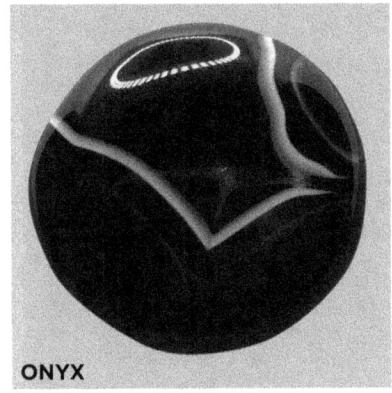

ONYX

ONYX is one specific variety of chalcedony that is typically black with white bands and occasionally brown bands. Onyx is often imitated by dying other stones black; usually agates which are a closely related variety of cryptocrystalline quartz. However dyed agates usually lack the distinctive bi-color banding and can have slight translucence which true onyx never has. True onyx usually displays a slightly waxy feeling and somewhat "flat", black luster as opposed to a bright shine, even when tumbled and smoothed. Slightly more common than black onyx, is sardonyx which is identical except that the general the colors occur in various shades of dark-red and sometimes with yellow banding, not white. As is true with onyx, much of the sardonyx on the market is dyed. Onyx is mined predominantly in Brazil, Pakistan and in a few locations in the Middle-East, and is far less common that most of other forms of chalcedony, making is slightly more costly on the market. Onyx jewelry and ornamentation has been used for several millennia.

Onyx is a stone of strength, power and stamina. It increases self-confidence and brings a feeling of ease and security in any surrounding. It helps overcome grief and emotional pain. Onyx helps empower self-control and stimulates wise decision-making. It encourages happiness and good fortune and good luck. Onyx treats disorders of the bones, bone marrow, blood cells and blood. It is also beneficial for teeth and the feet.

ASTROLOGICAL: *Gemini*, **CHAKRA***: Base, Element: Earth*

Related entries in this book: Agate, Chalcedony, Sardonyx

OPAL

FAMILY: Amorphous (Hydrated)
Quartz Silicate

CHEMISTRY: SiO_2nH_2O (Hydrated silica)

HARDNESS: 5.5-6

STREAK: White

SG: 1.98-2.25

STRUCTURE: Amorphous Rock

BLACK OPAL

OPAL is an amorphous, hydrated variety of silicate quartz chalcedony. Because of its lack of crystalline structure it is sometimes correctly classified as a mineraloid. Opals have a high water content of up to 20% but most commonly between 3% and 6%. Opals form as semi-transparent stones and also in many shades of subtle colors, the best examples displaying the typical opalescent play of light that makes them so popular as gemstones. Fire opal and black opal are treasured for their deep and varied bursts of colors. Specimens with well-defined spheres of multiple colors are considered precious gemstones. Opals are sensitive to changes in temperature and can chip or crack easily. Most of the world's opal supply currently comes from mines in a few locations throughout Australia. Recently, manufactured opals of superior gem-quality are being sold in the jewelry and precious stone market and are difficult to differentiate from true opal.

Opal is useful for cleansing and rejuvenating the body and spirit, and for support through emotional recovery. It has the ability to filter out useless information and focus the mind on the most relevant circumstances, bringing clarity and purpose to daily life. Physically, opal is a healer of fevers, chills and aches. It helps clear head colds and sinus infections.

ASTROLOGICAL: *Pisces,* **CHAKRA:** *Crown,* **ELEMENT:** *All*

Related entry in this book: Chalcedony

GREEN OPAL

PINK OPAL

OPALITE

FAMILY: Manufactured Glass

CHEMISTRY: SiO_2

HARDNESS: 5

STREAK: None

SG: 2.6

STRUCTURE: Amorphous Glass

OPALITE

THE OPALITE found in mineral shops, collections, and crystal healing arsenals, is a manufactured silica glass, with opal-like iridescence, that can include a visual color-shift from light blue to light yellow. Typically it is presented in the sizes and shapes of common rumbled stones or palm stones, or small figurines.

The term "opalite" is also used in the jewelry business to indicate milky-white, natural opal, that is used primarily for cabochons and rings. The two should never be confused. Polished stones found in rock shops and healing centers are invariably all of the manufactured variety; the natural opal being far too precious and costly to be sold in any large size.

Opalite should also not be confused with "opal glass" *(aka "milk glass")* which is a manufactured glass product made extensively throughout Europe in the 19th century, and used for high-end tableware and serving vessels. "Opal Glass" is also a trade name for products produced by Corning Glass works, et al. All of these are high-grade, decorative silicate glasses.

ASTROLOGICAL: *All,* **CHAKRA:** *Crown,* **ELEMENT:** *Air*

Related entries in this book: Goldstone, Obsidian Glass

ORPIMENT

FAMILY: Sulfide

CHEMISTRY: As_2S_3

HARDNESS: 1.5–2.0

STREAK: Yellow

SG: 3.49

STRUCTURE: Monoclinic

ARSENIC ORPIMENT

ORPIMENT is found near volcanic or hydrothermal and hot springs. It forms from the decay of the other common arsenate mineral realgar. It has a long history of use as a poison but also as a coloring agent and dye because of its beautiful orange or yellow hues. Ancient Romans and Chinese cultures used orpiment widely. It is highly toxic so it is rarely used for these purposes anymore. Caution should be used by collectors! Orpiment mineral is readily found for collecting but arsenates should be stored in sealed containers and if handled, thorough hand-washing is necessary.

As a healing crystal orpiment has been suggested to promote lucid dreaming and memory recall, however this author does not recommend using the arsenate minerals for healing because it may be unwise to handle them due to high toxicity. Please use caution!

ASTROLOGICAL: *Leo,* **CHAKRA:** *Solar Plexus,* **ELEMENT:** *Earth*

Related entry in this book: Realgar

PERIDOT

FAMILY: Nesosilicate

CHEMISTRY: $(Mg, Fe)_2SiO_4$

HARDNESS: 6.5–7.0

STREAK: White

SG: 3.20–4.30

STRUCTURE: Trigonal Orthorhombic Crystal

PERIDOT

PERIDOT is the gem quality, magnesium-saturated form of olivine. It is mostly found in igneous, deep mantle rocks and brought to the surface via volcanic activity. Peridot occurs in many shades of green and yellow, with the deeper green hues being the most desirable for gemstone use. In jewelry, the rich green peridot with flashes of yellow are considered to be finest quality. Even though it is not as durable as many other gemstones, recent finds in Burma, China and Pakistan have brought more high-quality stones to the market place and they have become quite popular. Until these finds, a very few locations in the Red Sea were historically the only mining source of peridot. Traces of peridot have also been found in meteorites but mostly as grain-sized specks, not of significant crystal size. Good specimens of peridot crystal clusters on a matrix of basalt are available at collector's shops for fair prices. They make excellent display pieces.

Peridot is a stone of protection and strength for perseverance in all challenging circumstances. Being a strong cleansing stone for the spirit, it opens the heart chakra and helps lift bad moods. Peridot is a stone that brings physical strength and stamina. It is also helpful when feeling stressed or depressed.

ASTROLOGICAL: *Leo,* **CHAKRA:** *Heart,* **ELEMENT:** *Air*

PETRIFIED WOOD

FAMILY: Cryptocrystalline Quartz

CHEMISTRY: SiO_2

HARDNESS: 7

STREAK: White

SG: 2.58-2.91

STRUCTURE: Trigonal Hexagon Crystal

PETRIFIED WOOD

PETRIFIED WOOD occurs as the result of a tree or plants having completely transitioned to stone by the process of pseudomorphism*. Organic wood materials from past ecosystems have been replaced over long periods of time, with minerals *(usually by a silicate, such as cryptocrystalline quartz)*, yet the original structure and textures of the plant is preserved intact. Chalcedony and onyx are common replacement minerals, making petrified wood quite hard. If the quartz replacement comes from a fairly pure silicate fluid, the color of the original wood can be preserved. However, more frequently, colored varieties of chalcedony bring vivid and rich hues to the process of petrification, making most specimens beautiful and desirable additions to any collection. Petrified wood is considered common and affordable, and can be found in just about any size. Due to its rigidity, it is often sliced and polished for display and certain specimens can resemble beautiful agate slices.

Petrified wood is good for grounding and stabilizing emotions. It is useful in calming primal fears, assists in nurturing one's practical side, and offers security to the bearer. It is a stone used to bring success in business, and is a good stone for general protection. Petrified wood calms the spirit and body and encourages contentment. Physically, it is beneficial for the bones, backaches, skin and hair.

ASTROLOGICAL: *None*, **CHAKRA:** *Heart*, **ELEMENT:** *Earth*

(pseudomorphism)—The process of one mineral replacing another mineral chemically, yet leaving the crystal structure, appearance, and usually the color intact.

PHOSPHOSIDERITE

FAMILY: Phosphates

CHEMISTRY: $FePO_4 2H_2O$

HARDNESS: 3.5–4.0

STREAK: White

SG: 2.72

STRUCTURE: Monoclinic Crystal

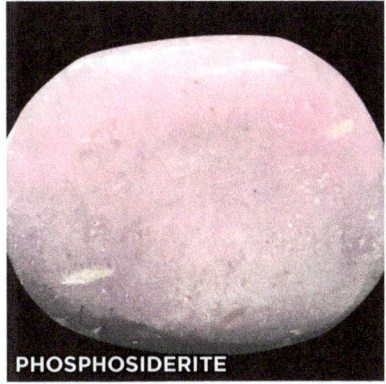
PHOSPHOSIDERITE

PHOSPHOSIDERITE is a fairly new mineral on the market and currently comes only from a few mine sites in Chile and Argentina. A rare mixture of phosphate and iron oxide, this stone displays beautiful hues ranging from lavender to purple, which is extremely unusual for a mineral containing iron. Being fairly soft, it is easily shaped into palm stones and ornaments, however it is probably too soft for use as a jewelry item.

Phosphosiderite is considered to be a stone of healing and hope. It also has the unusual and welcome benefit of addressing the specific needs of the bearer. Meditation with this stone may reveal solutions to problems that were buried in the subconscious. Being a highly energetic stone, phosphosiderite is useful for aligning the chakras and energy centers of the body.

ASTROLOGICAL: *None*, **CHAKRA:** *All*, **ELEMENT:** *Earth*

PIETERSITE

FAMILY: Cryptocrystalline Silicates

CHEMISTRY: SiO_2

HARDNESS: 7

STREAK: White

SG: 2.58–2.66

STRUCTURE: Trigonal Hexagon Crystal

PIETERSITE

PIETERSITE is an attractive stone in that it combines the colors and patterns of tiger's eye crystals, but has been broken, splintered and cemented back together over time by silicate quartz. Despite its brecciated *(mottled)* appearance, good specimens still display the typical chatoyance and play of light, like a tiger's eye. It fairly common commercially, and specimens can be found through any good source for collector's minerals. The past few decades have seen a great increase in the supply of blue pietersite. It has been received with enthusiasm by the jewelry market, and is priced accordingly. Small specimens can be quite costly, and pieces with the best blue colorings fetch prices comparable to some of the finer, traditional semi-precious gemstones.

Pietersite is generally considered to have the same metaphysical and healing uses as tiger's eye but due to its unusual mixed structure, it has unique and powerful properties. It is a useful companion that helps the bearer reach heightened awareness, stimulating the third eye and increasing intuition. Pietersite serves as a trusted guide during inner-reflection and helps to discover and recognize true purpose. Physically, it is suggested that it energizes the body, providing strength and stimulating the nervous system and brain. It helps with headaches, nervous system disorders, breathing difficulties, stomach pain and dizziness. In our family collection, we consider pietersite to be a "comfort stone", and have used it for multiple purposes.

ASTROLOGICAL: *None,* **CHAKRA:** *Third Eye,* **ELEMENT:** *Air*

BLUE PIETERSITE

PINOLITE (Peanut Stone)

FAMILY: Mixed Mineral

PINOLITE

PINOLITE (PEANUT STONE), also known as pinolith is an unusual and rare stone that makes an attractive addition to any collection. It is a rock of mixed minerals with a base of dolomite, blackened by inclusion of graphite, and sprinkled with "peanut-shaped" white inclusions of magnesite. It is found for collectors in masses or as cut slabs, but for collectors is tumble-polished in pebble to palm-sized specimens. It is soft but takes a polish well.

For healing purposes pinolite brings peace and relaxation, which makes it a good companion while trying to reach meditative states. Pinolite brings a calming energy to the bearer and releases nervous tension.

ASTROLOGICAL: *Capricorn,* **CHAKRA:** *Solar Plexus,* **ELEMENT:** *Water & Earth*

PREHNITE

FAMILY: Phyllosilicates

CHEMISTRY: $Ca_2Al(AlSi_3O_{10})(OH)_2$

HARDNESS: 6-6.5

STREAK: White

SG: 2.82-2.94

STRUCTURE: Orthorhombic Crystal

PREHNITE

PREHNITE is a phyllosilicate mineral with a combination of calcium and aluminum. It has an attractive and gentle transparency with a light green hue. Even though it is fairly brittle, it takes well to polishing, making it a beautiful semi-precious gemstone for jewelry. It has recently become more common in the commercial market for healers as well, often with the inclusion of dark, black streaks of epidote. Some specimens of prehnite in our own family collection display lacy flashes of gold when viewed in bright light. First discovered in South Africa by Dutch Colonel, Hendrik Von Prehn *(1733–1785)*, prehnite also occurs in Germany and Canada, which together are probably the main sources of most examples on the market today. Prehnite usually forms as botryoidal masses or stalagmitic lumps, and are quite stunning due to their color and transparency. Often, when individual botryoidal crystals are mined, they look strikingly similar to edible green grapes!

Prehnite is considered a stone of love and is a powerful crystal for healing the healer. It enhances premonition and reveals hidden knowledge. It enables one to always to be prepared for unexpected circumstances. Prehnite brings calm to the immediate surroundings and brings peace and protection to the bearer. Physically, prehnite is used to improve the general health of the kidneys, thymus, lungs and bladder. It relieves tension in the neck, shoulders and upper torso and is therapeutic for congested lungs due to colds or pneumonia. Prehnite also treats gout and blood disorders. Healers suggest prehnite to shrink malignant tumors and regenerate healthy cells.

ASTROLOGICAL: *Aquarius*, **CHAKRA:** *Sacral*, **ELEMENT:** *Water*

TUMBLED PREHNITE WITH EPIDOTE

PRESELI BLUESTONE
(Stonehenge Stone)

FAMILY: Igneous Rock

PRESELI BLUESTONE

PRESELI BLUESTONE is a common rock of rather ordinary igneous origin, with only slightly attractive visual appeal. Consisting of several minerals including dolerite, feldspar, biotite, quartz and others in minor amounts, Preseli Bluestone would ordinarily not be of much interest with one exception - it's unique and storied history. The famous Stonehenge circle in the Salisbury Plain in England is constructed entirely of huge monoliths of Preseli bluestone, making it one of the most famous stones in the world.

Of course the mystery of how 4000-8000 pound stones were moved 150 miles by an ancient civilization, from Preseli in Wales to Salisbury is a mystery, however it is no mystery how small specimens have recently made their way into the hands of rock shops and crystal healers - marketing! Tumbled palm stones and pebbles display a nice polish that enhances is slightly blue tint.

Healers have used the energies of bluestone to make connections to the deep past, to unlock past-life memories, and to open pathways to out-of-body experiences. It is considered to be one of the highest energy healing crystals and is becoming a treasured addition to many crystal healing collections.

ASTROLOGICAL: *Taurus,* **CHAKRA:** *All (Aura Crystals),* **ELEMENT:** *All*

PROPHECY STONE

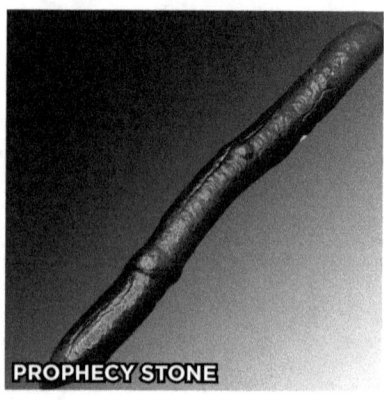

PROPHECY STONE

PROPHECY STONE is a very unusual concretion that can be composed of several different silicate-based minerals at its core, with a hard metallic exterior. A unique property of the "shell" is that is a pseudomorph mineral, meaning one mineral has been replaced over time by chemical means, but the physical and crystalline structure has not been altered. In most cases, prophecy stones display exteriors of marcasite (iron sulfate) structure that has been replaced by limonite (iron hydroxide). Most prophecy stones have been found in the Sharan Desert but the White Desert area of Egypt is the oldest known source for collectors. They form in a wide variety of shapes and sizes. Mist are palm-sized or smaller and can be star-shaped, rod-shaped, or amorphous.

As a healing crystal prophecy stones are thought to be very high vibration, radiating energy from the crown down through the entire body. As it's name suggests, prophecy stones are used for vision-seekers, clairvoyants, and psychics. They are said to be a great general healer for the entire body and spirit.

ASTROLOGICAL: *Sagittarius,* **CHAKRA:** *Crown/Third Eye,* **ELEMENT:** *Water*

Related entries in this book: Boji Stone™ , Moqui Marbles, Septarian

PURPURITE

FAMILY: Phosphates

CHEMISTRY: $Mn_3+(PO_4)$

HARDNESS: 4.0

STREAK: Purple

SG: 3.2

STRUCTURE: Orthorhombic

PURPURITE

PURPURITE is a brightly colored manganese and iron phosphate mineral that is becoming popular with collectors, as it becomes more available. It's beautiful purple hue obviously makes for some dazzling specimens! This mineral occurs commonly at zones of contact between manganese and iron, and sometimes forms only thin, colorful crusts, and at other times solid masses. Thus, specimens will have varied features. Purpurite does not typically take to polishing well, but it's natural condition is quite attractive without enhancement.

For crystal healing purposes purpurite is used for reaching into intuitive powers and connection to past-life experiences. Purpurite is a transitional stone used to assist with changing ones bad habits or altering harmful behaviors.

ASTROLOGICAL: *Virgo,* **CHAKRA:** *Crown,* **ELEMENT:** *Water*

PYRITE

FAMILY: Sulfides

CHEMISTRY: FeS_2

HARDNESS: 6–6.5

STREAK: Greenish-Brown/Black

SG: 4.95–5.10

STRUCTURE: Isometric Crystal

PYRITE CUBES IN MATRIX

PYRITE is iron sulfide and is usually found with other sulfides or oxides in both sedimentary and metamorphic rock formations, as well as in coal beds. It is the most abundant sulfide material found on Earth and has been used commercially since at least the 15th century as an ignition source for fire-arms. Pyrite is often found in quartz veins in association with gold and has a similar likeness due to small, glittering gold encrustations, earning it the popular nick-name, "fool's gold". Commercially, pyrite is a better source of sulfur dioxide than it is for iron, and the pyrite extracted currently in widespread use as a cathode material in lithium batteries. In certain locations, notably the Pyrenees mountains of Spain, perfect cubic crystals of pyrite grow in gypsum or clay/rock formations. These forms of pyrite cubes make a stunning display in any collection and are fairly abundant, inexpensive and easy to find.

Pyrite is of use to sharpen the mind and the intellect and enhance memory and retention of information. Pyrite assists in seeing behind facades and false-hoods, promoting insight into the words and actions of others. Physically, pyrite helps treat lung disorders and bronchitis and stops snoring. It can be used to reduce swelling and fevers and the associated aches and pains that come with high fever.

ASTROLOGICAL: *All,* **CHAKRA***: Crown,* **ELEMENT***: Fire*

Related entry in this book: Pyrite Sun

PYRITE FOOL'S GOLD

PYRITE SUN

FAMILY: Sulfides

CHEMISTRY: FeS_2

HARDNESS: 6-6.5

STREAK: Greenish-Brown/Black

SG: 4.95-5.10

STRUCTURE: Isometric Crystal

PYRITE SUN

PYRITE SUN is an unusual formation of iron sulfide that crystallizes after iron-bearing fluids permeate spaces between layers of shale and coal. These particular deposits are only found in the coal mining areas in Illinois and with the decrease in mining activity, they may not be found for very much longer. The estimated age of these formations is 350–300 million years ago during the Pennsylvanian period of the Cenozoic era, similar in time-frame to the world-famous fossil deposits near Mazon, Illinois. Quite substantial in size *(up to around 6 inches diameter)* and of hefty weight *(up to 5 pounds)*, they make a beautiful display piece for the collection. Pyrite suns are also known as pyrite dollars, and before modern mineralogical testing methods they were thought to be fossil remains of sea life such as sand dollars, or of plant life such as lily pads. Pyrite is a major source of iron ore, but because pyrite suns are associated with coal and not with iron mining sites, the majority of pyrite suns were collected by coal miners and local residents, and made their way to the mineral-collector's market. Good specimens are still easy to find for a decent price, and because of its shine, its size and its unique patterning, people are always drawn to the large, pyrite sun in our family collection. Pyrite, and all metallic mineral specimens, should be handled infrequently and kept moisture free. While oxidation is an inherent part of many attractive minerals, rust is not welcome on our metal specimens!

For healing purposes, pyrite suns are used for protection of all sorts and for increasing energy flow through the body. For best effect, the crystal is usually placed directly on the body during meditation or rest. Just as any iron mineral, pyrite suns are thought to help improve the health of the blood and circulatory system.

ASTROLOGICAL: *All*, **CHAKRA**: *Crown*, **ELEMENT:** *Fire*

Related entry in this book: Pyrite

QUARTZ

FAMILY: Quartz Silicates

CHEMISTRY: SiO_2

HARDNESS: 7.0 (Defining)

STREAK: White

SG: 2.71–2.73

STRUCTURE: Trigonal Hexagon Crystal

QUARTZ PRISMS

QUARTZ is almost pure silicate and is the second most abundant mineral on Earth's continental crust *(feldspar, also a silicate, being first)*. It can be found in veins with mineral ores, and is a component of granite and many other igneous rocks. Found almost worldwide, quartz varieties have been used for ornamentation since antiquity. Its internal crystal structure is hexagonal, yet quartz crystals can form individual, distinct prisms, or clusters of varying appearance, or as crusts and druse surfaces. Some of the finest, fully developed crystals are found in geodes, vugs and cavities in a wide-range of host materials. Clear, colorless quartz, called rock crystal, is valued for jewelry, carvings and ornament. Amethyst and citrine varieties are considered to be valuable semi-precious stones and are the purple and golden-yellow varieties of quartz. Quartz is also widespread in cryptocrystalline *(microscopic crystal)* forms that fall under the general description of chalcedony. This family includes agate, jasper, chrysoprase, onyx, heliotrope and carnelian. These can be found as separate entries in this book.

Clear quartz is known as the "master healer" and amplifies energy, as well as increasing the effectiveness of other crystals. It absorbs, stores, releases and regulates energy. Clear quartz draws off negative energy of all kinds. It provides focus and concentration, unlocks memories and helps clarify their meaning. Quartz stimulates and boosts the immune system, reduces fever, relieves headaches and dizziness and is a powerful healing tool in any aspect to which it is applied. Crystal healers and users of stones for spiritual purposes tend to "program" and maintain one special quartz crystal and treat it with special care, giving it an importance above all or most other crystals. A quartz crystal can become a strong and trusted ally for healing, and for the journey of self-exploration. Quartz crystals should be cleansed regularly in water, sunlight or moonlight.

ASTROLOGICAL: *All,* **CHAKRA***: Crown,* **ELEMENT:** *Fire*

Related entries in this book: Amethyst, Ametrine, Citrine

PHANTOM QUARTZ

CHLORITE PHANTOM QUARTZ

PHANTOM QUARTZ crystals form when crystal growth is interrupted and then begins to grow again. Each stage where growth has stopped is an opportunity for oxidation of the crystal surface or for different minerals to be deposited on the crystal. If crystal growth then resumes, these deposits remain as imbedded silhouettes that are the shape of the host crystal-faces. In the case of clear quartz the silhouettes form perfect, upward facing chevrons. This differs from inclusion quartz *(lodolite)*, or any other mineral that forms around a crystal of a different mineral, which becomes embedded when further growth continues around it. Chlorite deposits form beautiful green phantoms in quartz and they have recently become nearly as available as clear phantom quartz crystals. A good phantom crystal will retain great transparency and have multiple, well-formed chevrons. We are lucky to have a fair-sized and extremely clear phantom crystal in our family collection that clearly displays 7 distinct milky-white chevrons when viewed from any of the hexagon faces. It is astonishing to think of the immense passage of time between each stage of growth in this particular crystal and how it must have been completely undisturbed for countless ages to have remained in such pristine condition.

Phantom crystals prepare and guide the bearer for healing the spiritual self and uncovering deep lost feelings. It can detoxify the soul after negative energy experiences. They help to ground fluctuating feelings and emotions, making it possible to remain in balance and harmony with the world. These crystals are strongly attuned to the needs of the Earth and can bring healing energy to damaged environments and wildlife. Phantom crystals are stones of deep knowledge and wisdom and are trusted guides to help remember past life experience and to access the Akashic records. Phantom quartz crystals are purposeful for improving any disease and ensuring that the vitamins and minerals ingested are incorporated fully into the body.

ASTROLOGICAL: *All,* **CHAKRA:** *Crown,* **ELEMENT:** *Air*

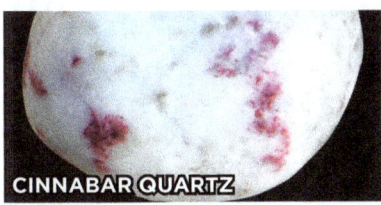

BLACK QUARTZ is a powerful detoxifier and anti-oxidant. It offers spiritual clarity and protection from negative forces that may distract from one's true path. The black crystals are due to non-oxidized iron invading the geode during crystal growth and makes quite a mesmerizing appearance.

ASTROLOGICAL: *Pisces,* **CHAKRA:** *Heart,* **ELEMENT:** *Earth*

BLUE QUARTZ is rarely found in complete prisms or transparent form, but can be found easily in rough, massive form. Color can range from pale to very deep, rich blue, the later usually being the most desirable. It is often mistaken for sodalite or lazurite and can be distinguished by its superior hardness.

ASTROLOGICAL: *Virgo,* **CHAKRA:** *Throat,* **ELEMENT:** *Water*

CINNABAR QUARTZ is a milky quartz with trapped "globules" of shocking-red, cinnabar mercury. Even though the mercury is encased in stone, caution should be used handling this mineral. **Please keep away from children and wash your hands after touching.**

ASTROLOGICAL: *All,* **CHAKRA:** *Sacral,* **ELEMENT:** *Fire*

GREEN QUARTZ

FUCHSITE QUARTZ

LODOLITE QUARTZ

RED QUARTZ

GREEN QUARTZ forges unity, honesty, community spirit and tolerance. It is a stabilizer for the endocrine system and adrenal glands, and used for ailments and diseases of the heart and lungs and for steadying heart palpitations and arrhythmia.

ASTROLOGICAL: *All,* **CHAKRA:** *Crown,* **ELEMENT:** *Air*

FUCHSITE QUARTZ is fairly rare. There are numerous other minerals to be found in quartz, since silicate solutions are so abundant in mineral deposits. Together, they are all known as "inclusion quartz" or "lodolite". *(See the separate listing for lodolite quartz).*

ASTROLOGICAL: *Aries,* **CHAKRA:** *Solar Plexus,* **ELEMENT:** *Air*

LODOLITE QUARTZ (LODALITE) is mineralogically, any quartz with visible inclusions of other minerals. It is also referred to as "fantasy quartz". Metaphysically, lodolite quartz is a very high-vibration stone used for astral projection, past-life seeking and exploring spiritual realms.

ASTROLOGICAL: *All,* **CHAKRA:** *Third Eye,* **ELEMENT:** *Earth*

RED QUARTZ has a layer of oxidized iron ions included in the outer layers of the crystal. This probably occurred due to an iron-rich, silicate solution invading the site where the crystal was growing. Spiritually, red quartz is used to enhance meditation and focus on the inner-self.

ASTROLOGICAL: *All,* **CHAKRA:** *Sacral,* **ELEMENT:** *Fire*

ROSE QUARTZ is the stone of unconditional love. An obvious stone for opening the heart chakra, it also opens the spirit to accepting love from others. Healers use rose quartz to lower blood pressure by reducing stress and anxiety. Is helpful for children with bed-wetting problems and it brings clarity of mind while studying or trying to retain new information.

ASTROLOGICAL: *All*, **CHAKRA:** *Crown,* **ELEMENT:** *Air*

SMOKY QUARTZ helps to ground the spirit and induces energy flow from the higher chakras to the lower chakras. It helps with finding a balance between physical needs and spiritual desires. Smoky quartz is a good crystal to use for pain relief and general healing.

ASTROLOGICAL: *All*, **CHAKRA:** *Base,* **ELEMENT:** *Earth*

SPIRIT QUARTZ is an example of several stages of crystal growth probably due to changing conditions of the fluid present, and/or changes in pressure and temperature. The large quartz-prism crystals stopped growing, and at some point a coating of much smaller crystals grew, creating a "druse" surface that sparkles and glitters. There is currently plenty of spirit quartz available and it is a wonderful addition to any collection. The spirit quartz in the photo above, is our own wonderful specimen that shows well-defined and terminated quartz prisms, coated with a druse crust of small crystals of quartz, amethyst and citrine all in the same cluster. It's an amazing piece and certainly, one of our very favorites in the collection!

ASTROLOGICAL: *None,* **CHAKRA:** *All,* **ELEMENT:** *Water*

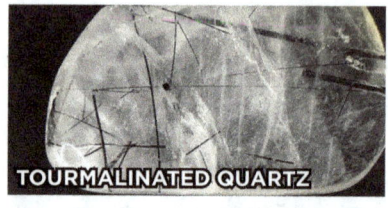

TANGERINE QUARTZ

TOURMALINATED QUARTZ

TIBETAN QUARTZ

TANGERINE QUARTZ helps to find clarity in the chaos of the world, allowing recognition of wisdom and truth. It raises the body's vibration level and makes interactions with others more enjoyable and productive. Physically, tangerine quartz helps to speed recovery from stomach flu and reduces aches and pains from fever. Also, it helps the skin and tissues heal from bruises, cuts and scrapes.

ASTROLOGICAL: *All,* **CHAKRA:** *Sacral,* **ELEMENT:** *Fire*

TIBETAN QUARTZ has inclusions of carbon and hematite or goethite that give it a rich, dark color. Tibetan quartz is used to access ancient wisdom and spiritual messages. It helps one to get to the source of a problem so it can be healed. It can be used to form bridges between past life experiences and present life. Tibetan quartz is a powerful healing crystal that aids with recovery from any illness or disease and is a powerful detoxifier and anti-oxidant. Clarification: Not all Tibetan quartz comes from Tibet. Many specimens come from China, India and Pakistan.

ASTROLOGICAL: *All,* **CHAKRA:** *Crown,* **ELEMENT:** *Fire*

TOURMALINATED QUARTZ has inclusions of black tourmaline and is a powerful crystal with strong grounding energy and is a potent protector. It brings heightened awareness and an increase in psychic energy. Physically, tourmalinated quartz provides protection from environmental pollutants and electromagnetic radiation. Good for relieving eye stress due to extended use of computers, television and home electronics.

ASTROLOGICAL: *All,* **CHAKRA:** *Crown,* **ELEMENT:** *Fire*

RUTILATED QUARTZ

FAMILY: Oxides

CHEMISTRY: TiO_2 & SiO_2

HARDNESS: 6.5

STREAK: Red

SG: 4.23–4.25

STRUCTURE: Tetragonal Crystal

RUTILATED QUARTZ

RUTILATED QUARTZ includes hairs or rods of titanium dioxide. It can occasionally be found as an inclusion in quartz, although the quantity and quality of rutile inclusions vary greatly as does the clarity of the quartz. Rutile forms beautiful, needle-like crystals with golden or sometimes silvery, or black, metallic sheen. Rutile is the main source of titanium used for industry. The primary source of rutile in quartz is currently Sierra Leone, in Africa. Some smaller deposits of extremely high quality can be found in Switzerland, with needles of rutile exceeding 12 inches in length. We have some small raw specimens of rutilated quartz with little or no clarity in our family collection, yet they still display fascinating sprays of rutile needles on their surfaces. However, we do have several, significantly clearer, natural quartz prisms displaying internal rutile needles, that are among our most treasured specimens of any type. The suspension of the perfectly linear and incredibly delicate rutile crystals "trapped" in a piece of perfectly clear quartz, can be quite enchanting. I am known to be carrying any number of stones in my pockets at all times, and there's a good chance that one of them will always be a rutilated quartz. I am quite in love with it.

Rutile defends against negative energy and protects against disease. It helps the bearer to solve "unsolvable" problems. It brightens moods, lifts depression and relieves anxiety. When rutile is included in quartz, they amplify each other's energy.

ASTROLOGICAL: *Gemini*, **CHAKRA:** *All*, **ELEMENT:** *Earth*

RUTILATED QUARTZ

RAINFOREST RHYOLITE

FAMILY: Igneous, Felsic Rock

CHEMISTRY: Mixed Mineral

HARDNESS: 6.5–7

STREAK: White

SG: 2.41–2.62

STRUCTURE: Trigonal Hexagon Crystal

RAINFOREST RHYOLITE

RAINFOREST RHYOLITE is igneous in origin and is technically a feldspar rhyolite or granite mineral. While most granites usually form as macroscopic crystals, rainforest rhyolite probably cooled under more pressure during formation, making it a cryptocrystalline rock similar in structure, but not in composition, to jasper. Much the same as leopardskin jasper, rainforest rhyolite may be as much as 65% alkali and plagioclase feldspars which are silicates, but not quartz. Rainforest rhyolite has traditionally been described and sold as a jasper, however recently it appears more often with its proper classification as a rhyolite *(unlike leopardskin jasper which still carries its misnomer)*. Rainforest rhyolite comes primarily from Australia near an extinct volcanic zone and was probably formed deep in lava flows where pressure created gas bubbles in the lava, which are responsible for the spherical inclusions *(biotite and other minerals)* in the stone. Similar looking rhyolites *(and jaspers)* can be found in China, Brazil and Madagascar. The name "rainforest" is given mostly to represent its lush, green hues and is probably not due to geographic association with any rainforest location.

Rainforest rhyolite is a great balancing stone that helps with issues of self-esteem and emotional stability. It brings a sense of strength and perseverance to face issues that need addressing. Rainforest rhyolite is of great use in reducing fear and anxiety. It is helpful in expressing personal insight and finding creative ways to solve problems. Healers suggest using rainforest rhyolite to boost the immune system, and to cleanse the body's systems and internal organs, most specifically, the kidneys, spleen and liver.

ASTROLOGICAL: *Sagittarius,* **CHAKRA:** *Heart,* **ELEMENT:** *Fire*

Related entries in this book: Leopardskin Jasper

REALGAR

FAMILY: Sulfide

CHEMISTRY: As_4S_4

HARDNESS: 1.5–2.0

STREAK: Orange

SG: 3.56

STRUCTURE: Monoclinic

ARSENIC REALGAR

REALGAR is one of the two main arsenate minerals; orpiment being the other. Same as orpiment was used in ancient times as a deep yellow pigment, so was realgar used for its ruby red color. Realgar was also commonly used as a poison for killing weeds and pests such as rats. It is typically formed near hot springs and hydrothermal vents such as geysers, and also as a minor mineral in lead, silver, and gold ores. It is possible to find attractive small crystal crusts of beautiful, red realgar, sometimes in association with yellow orpiment, but remember it is highly toxic and should be kept in sealed containers. Hand-washing after handling immediately after is a must.

   For crystal healing purposes realgar is said to be a calming influence, reducing rage tendencies and bringing more focus to inner-peace and stability. However, this author does not recommend handling realgar for healing purposes due to its extreme toxicity. **Please exercise caution!**

ASTROLOGICAL: *Gemini*, **CHAKRA:** *Solar Plexus*, **ELEMENT:** *Earth*

Related entry in this book: Orpiment

RHODOCROSITE

FAMILY: Carbonates

CHEMISTRY: $MnCO_3$

HARDNESS: 3.5-4

STREAK: White

SG: 3.51-3.70

STRUCTURE: Trigonal Hexagon Crystal

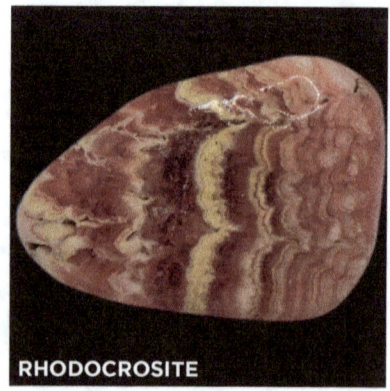

RHODOCROSITE

RHODOCROSITE is a manganese carbonate mineral with gentle pink hues and interspersed with bands of white. It commonly occurs in hydrothermal zones and is frequently associated in veins when mining for silver. Argentina and Romania are currently the primary sources for rhodocrosite mining. It is a major source of manganese ore, which has commercial uses in the manufacture of stainless steel and aluminum alloys. The finest examples of rhodocrosite used for jewelry and decorative purposes are taken from stalactites and botryoidal masses formed in caverns. These formations display characteristic, concentric bands and waves which are most attractive and more pronounced, when sliced or polished. Because of its profound beauty, rhodocrosite is one of the few soft minerals that is commonly used for rings and pendants even though it could potentially chip and break. Fine specimens used for jewelry are considered semi-precious and are priced accordingly. Lesser quality samples can be purchased at most collector's shops and at some healing crystal stores.

Rhodochrosite is an effective balancing crystal for both physical and spiritual energy. It promotes feelings of love and passion while energizing the spirit. It opens the heart, lifts depression and encourages a positive and cheerful outlook. Physically, rhodocrosite is used to treat migraines, skin disorders, thyroid imbalances, and intestinal problems. It helps regulate heartbeat and reduces high blood pressure. It can help children with bed wetting problems.

ASTROLOGICAL: *Leo,* **CHAKRA:** *Heart,* **ELEMENT:** *Earth*

RHODONITE

FAMILY: Inosilicates

CHEMISTRY: $MnSiO_3$

HARDNESS: 5.5–6.5

STREAK: White

SG: 3.40-3.70

STRUCTURE: Triclinic Crystal

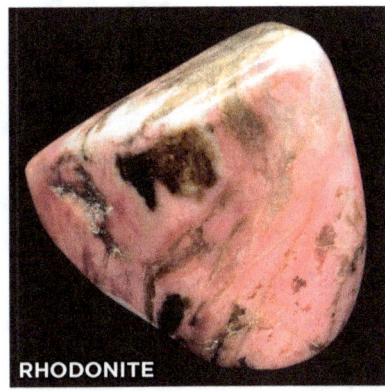
RHODONITE

RHODONITE is a rich pink, pyroxine inosilicate mineral, and like rhodochrosite, it takes its name from the Greek word "rhodon" which means rose. Large, gem-quality crystals of Rhodonite are uncommon but massive formations of rhodonite display beautiful contrast between pink manganese, black manganese oxide *(sometimes replaced by zinc)* and white calcium and silica. It is typically not hard enough to be popular as a cut gemstone, yet it tumble-polishes fairly well, which greatly enhances the intensity its colors. We have also had good luck at home cutting raw rhodonite into palm stone shapes and then polishing by hand. Raw or tumbled examples of rhodonite are inexpensive and readily available to collectors. Rhodonite is the official state mineral of Massachusetts.

Rhodonite opens the heart and balances the chakras making it useful for recovering from emotional stress. It is a grounding stone for the spirit, balancing the male and female yin/yang. It relieves stress and brings calm nerves and confidence. Physically, rhodonite is suggested for soothing itches and insect bites. It helps heal wounds with minimal scarring, and has long been a favorite crystal for bringing fertility. Rhodonite is used by healers to treat maladies of the lungs such as emphysema and pneumonia, as well as auto-immune diseases and stomach ulcers. Healers frequently suggest rhodonite for reducing inflammation of the joints and for arthritis, especially in the hands and feet.

ASTROLOGICAL: *Taurus*, **CHAKRA:** *Heart*, **ELEMENT:** *Earth*

RUBY IN FUCHSITE

FAMILY: Oxide in Phylosilicate

CHEMISTRY: $K(Al,Cr)_2(Si_3AlO_{10})(OH)_2$

HARDNESS: 2-4

STREAK: White

SG: 2.60. -3.01

STRUCTURE: Orthorhombic Crystal

RUBY IN FUCHSITE

RUBY IN FUCHSITE Rubies can occasionally be found in association with fuchsite within pegmatite intrusions where micas are present. A rock with inclusions of both minerals is commonly referred to as "ruby fuchsite". Fuchsite is a pale green variety of muscovite mica *(colored by chromium impurities)*. Ruby is red corundum *(colored red by aluminum and chromium impurities)*. Each inclusion of ruby is usually ringed by green tourmaline, which is a much darker green than the fuchsite content of the rock. Ruby in fuchsite has become very popular on the collector and spiritual healing markets since the early part of this century, when exports from India increased dramatically. South Africa is still a good source for some of the finer specimens on the market. Ruby fuchsite is often labelled or sold as ruby in zoisite (and visa versa). There is much debate regarding differentiating the two. In general, ruby fuchsite is usually slightly paler green and the ruby inclusions have the characteristic rings of green tourmaline, while ruby in zoisite does not have rings and is darker green overall. Also, zoisite more often occurs in metamorphic rocks, not pegmatites in igneous rock like fuchsite does, but the consumer has no way of making a determination about the rock's origin.

Ruby fuchsite helps clear blockages of the heart chakra and brings gentle and positive energy to the bearer. It helps with self-awareness, while opening connection to others. It is a great comfort stone when grieving. It helps increase physical energy, which is helpful when recovering from extended or chronic illness. Ruby fuchsite strengthens the heart, and treats circulation issues. It is a great assistant for those who have trouble falling asleep and staying asleep soundly.

ASTROLOGICAL: *Aquarius,* **CHAKRA:** *Heart,* **ELEMENT:** *Air*

Related entries in this book: Fuchsite, Ruby

SARDONYX

FAMILY: Cryptocrystalline Silicate

CHEMISTRY: SiO_2

HARDNESS: 7.0

STREAK: White

SG: 2.84–2.88

STRUCTURE: Trigonal Hexagon Crystal

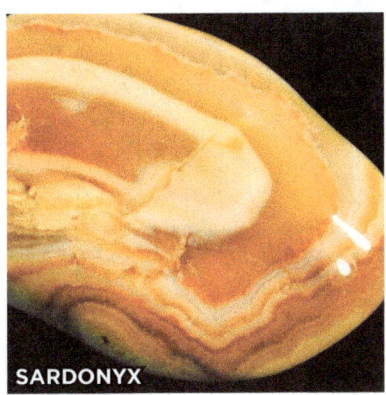

SARDONYX

SARDONYX is a variety of chalcedony that is identical to black onyx except that sardonyx has impurities of iron oxide, which are responsible for its reddish-brown hues. Sardonyx, like onyx, is often imitated by dying agates, which are a closely related variety of cryptocrystalline quartz. Dyed agates usually lack the distinctive contrast in the banding and can have slight translucence which true sardonyx *(and onyx)* never has. Slightly more common than black onyx, sardonyx can be found in several locations around the world, including India, Russia, United States, and Peru.

Sardonyx has many of the same metaphysical characteristics as onyx. It is a stone of strength, power and stamina. It increases self-confidence and a sense of security in any situation. It helps overcome grief and emotional pain. Sardonyx empowers the bearer with self-control and stimulates the power of wise decision-making. It encourages happiness and good fortune. For physical healing purposes, sardonyx treats disorders of the bones, bone-marrow and blood. It is beneficial for healthy teeth and it rejuvenates tired and sore feet.

ASTROLOGICAL: *Gemini,* **CHAKRA:** *Base,* **ELEMENT:** *Earth*

Related entries in this book: Chalcedony, Onyx

SCHALENBLENDE

FAMILY: Mixed Mineral

SCHALENBLENDE

SCHALENBLENDE is an unusual and very unique conglomerate of the minerals marcasite, wurtzite galena, sphalerite, and possibly others. This mostly metallic mineral forms as hydrothermal deposits, concretions, and stalagmites, and as such displays mixed and highly textured patterns of alternating metals and rock. The metals in the mineral display beautifully when a specimen is cut or planed flat and polished.

Schalenblende is rarely found in the crystal healing world, possibly because of scarcity of specimens and also since specimens that are available usually end up in mineral collectors cabinets. Schalenblende concretions are also fairly brittle and tarnish easily, they are not commonly handled.

ASTROLOGICAL: *Capricorn,* **CHAKRA:** *Base & Third Eye,* **ELEMENT:** *Earth*

Related entries in this book: Galena, Marcasite, Sphalerite

SCOLECITE

FAMILY: Tectosilicate Zeolites

CHEMISTRY: $CaAl_2Si_3O_{10}\ 3H_2O$

HARDNESS: 5–5.5

STREAK: White

SG: 2.16–2.40

STRUCTURE: Monoclinic Crystal

SCOLECITE

SCOLECITE usually forms as lustrous, milky-white, masses with well-defined striations or translucent veins. When polished, this creates a ghostly striated effect. In some cases Scolecite is found as thin, dramatic sprays of prismatic needles. In this crystalline form it is fairly uncommon, yet is the most desirable for collectors of exotic minerals. It tends to occur along with other zeolites in hydrothermal veins and in association with igneous, volcanic rocks such as basalt. Scolecite and other zeolites are used commercially for filtration of gasses, water and other liquids. Zeolites have a permeable molecular structure which allows the passage of certain fluids or gasses while filtering-out undesirable contaminants. Scolecite is one of the widest-ranging zeolites, being found in such diverse places such as Brazil, India, California, USA and Iceland.

Scolecite can be of great use to bring peace and spiritual transformation. It connects one heart to another and strengthens bonds. Scolecite helps with relaxation, brings tranquility and overcomes sleeping disorders. It is a stone of connection and harmony which removes blockages and brings inner peace. Physically, scolecite helps maintain proper serotonin levels in the brain.

ASTROLOGICAL: *Capricorn,* **CHAKRA:** *Third Eye,* **ELEMENT***: Air*

Related entries in this book: Cavansite, Stilbite

SELENITE

FAMILY: Sulfates

CHEMISTRY: $CaSO_4$ $2H_2O$

HARDNESS: 2.0

STREAK: White

SG: 2.29–2.31

STRUCTURE: Monoclinic Crystal

SELENITE

SELENITE is a crystalline form of gypsum. It is naturally transparent but can appear milky or even opaque with the inclusion of impurities that were present as the crystals grew. Selenite crystals can be found worldwide and usually occur in evaporate deposits where non-crystalline, massive gypsum is present and can grow to impressive lengths. In the year 2000 workers at the Naica lead and silver mine in Chihuahua, Mexico discovered a large, limestone cave at a depth of over 1000 feet. Within the cave are massive, selenite crystals over 40 feet long and 5 feet wide that weigh approximately 50 tons each! Geologists have determined that these amazing crystals probably formed in an underwater environment rich in anhydrite, which dissolved into selenite. Selenite, gypsum and anhydrite are chemically identical, except anhydrite lacks water molecules in its chemistry. When it is hydrated it becomes gypsum. (*The same can happen to anhydrite specimens in a mineral collection, so precautions must be taken to keep them dry*). Selenite is abundant, inexpensive and attractive in any collection. Some common forms sold, are crystalline rods and towers. Carved shapes, such as hearts and palm stones, and small tumbled or polished stones should be handled with care because of the brittle nature of selenite. Polished selenite displays a beautiful translucence and a unique depth of light showing inner-glow and surprising chatoyance.

Selenite is very energetic and can help open pathways to higher levels of consciousness. It purifies and recharges other crystals and removes negative energy from them, without needing cleansing itself. Selenite can be used to purify rooms and to protect public places. Selenite reduces stress and anxiety, bringing calm and clarity to the mind, and relaxation to the body.

ASTROLOGICAL: *Gemini*, **CHAKRA:** *Crown*, **ELEMENT:** *Air*

Related entries in this book: Anhydrite, Gypsum Desert Rose

SEPTARIAN

FAMILY: Carbonate & Silicate Concretion

CHEMISTRY: Mixed Mineral

HARDNESS: Varies

STREAK: White

SG: Varies

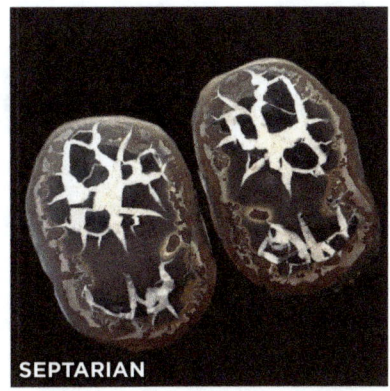

SEPTARIAN

SEPTARIAN nodules display a consist of yellow calcite, gray bentonite, brown aragonite and clear or white barite. The formation of these concretions is not entirely understood, but generally they are found in beds where alternating shallow sea water and dry evaporate zones occurred repeatedly during the Cretaceous period 50 to 70 million years ago. These nodules formed as bentonite-rich mud from the sea was tumbled into loose balls by wave or tidal action. Due to evaporation they were exposed to air and dried and cracked. When submerged again, calcium carbonate from decomposing sea life was deposited in the cracks and they filled in with calcite material. Upon exposure to atmosphere yet again, aragonite crystals formed from chemical interaction between the calcite and bentonite. This gives a septarian nodule its typical appearance of three different minerals. When sliced, the cross-section of the concretion can display a beautiful "cracked geode" effect that often exposes internal barite crystals. Polished septarian nodules have become fairly common in the collector's market and slices of septarian even more-so. Most slices have had a transparent glue or binder applied to keep the delicate interior calcite and barite crystals intact. Purists may consider this an alteration and an unnatural addition, however it is a practical and necessary process to ensure that a fine and delicate specimen is not damaged.

Septarian is not a very common tool in the spiritual healing world, but is becoming increasingly popular as more easily-handled, polished and tumbled nodules enter the retail and collector's markets. Most sources agree that septarian is a stone of high-vibration that creates clarity and self-awareness. This can be useful when preparing for public events and it facilitates composure when speaking or performing in public. Septarian slices are extremely captivating to gaze at and can be a nice transitional stone when exiting meditational states.

ASTROLOGICAL: *Taurus,* **CHAKRA:** *Third Eye,* **ELEMENT:** *Earth*

Related entries in this book: Boji Stone™ , Moqui Marbles, Prophecy Stone

SERAPHINITE

FAMILY: Phyllosilicates

CHEMISTRY: $(Mg,Fe,Al)_6(Si,Al)_4O_{10}(OH)_8$

HARDNESS: 3.5

STREAK: White

SG: 2.6–3.0

STRUCTURE: Monoclinic Crystal

SERAPHINITE

SERAPHINITE is a complex magnesium and iron silicate mineral, and one of a group of similar minerals called chlorides. A high degree of chloride molecules in its chemistry give seraphinite its rich and varied green hues. The many "feathery"and semi-translucent crystals in this stone have been warped, stretched and compressed due to metamorphic pressure which creates a beautiful, wispy pattern and texture throughout the stone. Fine examples of seraphinite are used for jewelry, usually in cabochon form, to enhance the depth of pattern. For such uses, the stone is usually backed with metal because it is a very soft and not durable for many jewelry purposes. Seraphinite occurs in only one site, near an iron ore deposit still being mined in Russia. It is mined under the mineral name "chlinichlore", in several locations in South America but the specimens are rarely considered to be jewelry-grade. Seraphinite is sometimes referred to by the trade name "chlorite jade".

Healers find this stone to be extremely high-energy, being full of light and love, and use it for many purposes. It is known as the "angel stone" and is associated with, higher energies, spirit guides and heightened states of being. Seraphinite can boost the effectiveness of any meditative, spiritual or psychic endeavor, and is an aid to all forms of physical healing.

ASTROLOGICAL: *Sagittarius,* **CHAKRA:** *Heart,* **ELEMENT:** *Earth*

SERPENTINITE

FAMILY: Phylosilicates

CHEMISTRY: $[(Mg, Fe)_3Si_2O_5(OH)_4]$

HARDNESS: 2.5–5.5

STREAK: White

SG: 2.50–2.65

STRUCTURE: Hexagonal Crystal

SERPENTINITE

SERPENTINITE is a mineralogical and geological term that refers to any of 20 varieties belonging to a polymorphic* mineral group found in a number of metamorphic rock formations. Serpentines are metamorphic rocks formed at relatively low temperatures by chemical processes of oxidation and hydration in the presence of magnesium, iron and other minerals mixed with water. Serpentinite is one member of this group and is almost always found in mixed, fragmented shades of green due to alternating layers of cryptocrystalline phyllosilicates and magnesium hydroxides *(Mg(OH)2)*. Subjected to further cycles of weathering, hydration and oxidation, some serpentine minerals will become the attractive, purple mineral, stichtite *(see separate listing in this book for stichtite)*. Due to the random mixing and folding of these layers, and similar mineral compositions, serpentine varieties are not easy to differentiate. Serpentinite is brittle due to its layering, yet is tough enough to carve or polish carefully. Some specimens that have a less mottled appearance and gentle green hues, have been used as a replacement for the much more costly minerals, jadeite and jade nephrite. The specimen shown in the photo above, is the variety known as "lizardite" due to its scaly, green appearance.

Serpentinite assists with accessing the records of Akashic wisdom and knowledge, and helps to retrieve memory of past lives. It clears the chakras and opens pathways for kundalini energy. Serpentinite is good for treating conditions of the skin, and complexion. It brings relief from moon cycles and cramps and is beneficial to the health of the heart and lungs.

ASTROLOGICAL: *Gemini,* **CHAKRA:** *Heart,* **ELEMENT:** *Fire*

Related entry in this book: Stichtite

*(polymorphic)—Having the exact same chemistry but a different crystal structure.

SHIVA LINGAM

FAMILY: Cryptocrystalline Quartz

CHEMISTRY: SiO_3

HARDNESS: 3.5–4

STREAK: White

SG: 2.70–2.66

STRUCTURE: Trigonal Hexagon Crystal

SHIVA LINGAM

SHIVA LINGAM stones are traditionally harvested only from the Narmada River in India, although they are also extensively produced artificially by tumbling and polishing techniques. Natural lingams are formed over time from the natural inflow and outflow of the river which has enough current to tumble the stones forward and back on the river bed, but without enough energy to lift the stones and change their orientation. This produces the typical egg-shape, instead of a rounded pebble which would be found in a high-flow, high-energy water system or tidal zone. Lingams are composed mainly of chalcedony *(cryptocrystalline quartz)* with traces of iron. It is believed that the area around the Narmada River was impacted millions of years ago by a large meteor which spread iron and nickel throughout the region. This is possibly the cause of the dark banding which occurs frequently in the lingams, and is also reason for the belief in the high spiritual powers of the stone, since some of its composition comes "from the stars". Shiva lingams are a common item at healing and spiritual shops, but are also a fine specimen of banded chalcedony for those interested solely in mineralogical collecting.

In Hinduism, shiva lingams are one of the most sacred formations in the world and have been used for worship for more than three millennia. It is the holy symbol of Lord Shiva and represents the sacred aspects of male and female fertility and is a symbol of the creation of life. A shiva lingam can charge the entire chakra system, activate kundalini energies and enhance inner-transformation. It can break up old life patterns and direct the bearer on a new journey through life. It promotes strong feelings of unity between the self and others, even when separated by great distance or time. Shiva lingam has traditionally been used as a tool for infertility and impotence.

ASTROLOGICAL: *Scorpio,* **CHAKRA:** *Base,* **ELEMENT:** *All*

Related entry in this book: Chalcedony

SHUNGITE

FAMILY: Metamorphic Carbon

CHEMISTRY: Variable

HARDNESS: 3.5–3.6

STREAK: Black

SG: 2.10–2.20

STRUCTURE: Amorphous

SHUNGITE

SHUNGITE is a fossilized organic carbon material composed of Precambrian sea sediments. Shungite has a high carbon content *(30%)*, and contains silica *(45%)* and mica materials *(20%)*. The remaining content of shungite being made up of small and varying amounts of metallic elements that may have precipitated into the stone, after the organic deposits had metamorphosed. A higher degree of metamorphism is what distinguishes shungite from coal. It is named for the Shunga region of Russia where it was discovered and where it is still mined extensively for industry. It is used commercially for use in filtration applications and as an additive to carbon coke in blast furnaces. Shungite contains some carbon fullerenes* and has possible implications for future use in nano-technologies.

Shungite is a very protective stone, shielding the bearer from malevolent spiritual energies, psychic attacks and negativity. Small pieces of shungite have recently become widely available at healing shops and at rock and mineral retailers. Because of its high carbon content, it has been suggested to stir a few pieces into a glass of drinking water to purify it, or to use the cleansed water as a mouth rinse. **I do not recommend that you ingest any mineral, in any form unless instructed by a physician or licensed clinician, and are ensured of the purity and safety of the product!**

ASTROLOGICAL: *None,* **CHAKRA:** *Root,* **ELEMENT:** *Earth*

*A fullerene is a molecular, carbon structure, similar to graphite but with the unique property of its atoms forming into balls or tubes. The structure of these molecules looks so similar to the designs of Buckminster Fuller's geodesic dome architecture, that it was named for him by its discoverers at Rice University in 1985.

SODALITE

FAMILY: Feldspathoid Tectosilicates

CHEMISTRY: $Na_8(Al_6Si_6O_{24})C_{l2}$

HARDNESS: 5.5–6

STREAK: White

SG: 2.27–3.33

STRUCTURE: Isometric Crystal

SODALITE

SODALITE is a lightweight, fairly hard yet fragile mineral. It is named after its sodium content. It is classed mineralogically as a feldspathoid which means it has a much lower silicate content than a proper feldspar even though it may be of similar chemistry otherwise. Sodalite can display a variety of colors, most commonly deep blue that can be confused with lazurite *(lapis lazuli)*. Lazurite has a similar chemistry except it lacks the chlorine molecule that sodalite has, and is replaced instead with sulfur sulfate. Sodalite is indeed one component of lapis lazuli, and the main method of differentiating these minerals is that sodalite almost always has inclusions of white or colorless material that lazurite rarely has and sodalite lacks the shiny, pyrite inclusions that are a distinguishing feature of lapis. Sodalite usually forms in silicate-poor volcanic intrusions and is found throughout South America, Canada and the United States. It has become common on the commercial and retail market in both raw and polished forms and makes a colorful addition to any rock collection. It cuts, tumbles and polishes well, but is generally too soft to be used as jewelry.

Sodalite helps bring a sense of order and calm to the mind. It encourages rational thought and intuition. It helps the bearer to both discover, and verbalize difficult inner feelings. It can balance emotions, calm panic attacks, and reverse bad moods and crabbiness. Sodalite is great stone to keep handy for children! Sodalite boosts the general metabolism and immune system, and helps the body absorb calcium. It treats the throat, vocal cords and larynx and helps with hoarseness. It has been recommended to cool a fever, lower blood pressure, and to alleviate dehydration. Sodalite is suggested as a cure for insomnia, and helps children with bed wetting problems.

ASTROLOGICAL: *Sagittarius*, **CHAKRA:** *Throat*, **ELEMENT:** *Water*

Related entry in this book: Lazurite (Lapis Lazuli)

SPINEL

FAMILY: Oxides

CHEMISTRY: $MgAl_2O_4$

HARDNESS: 7.5–8

STREAK: Gray

SG: 3.60–4.10

STRUCTURE: Isometric Crystal

SPINEL

SPINEL is a fine, precious gemstone and occurs in almost any color. Red spinel is the most precious and sought-after variety. A good spinel crystal can rival ruby for color and luster. In fact, many famous rubies throughout history have, upon modern inspection methods, proved to actually be spinels. It is a fairly abundant mineral and the use of spinel for jewelry is once again growing in popularity, after being somewhat unfashionable for a long time due to spinel being portrayed as "poor man's rubies". A good example of how sentiment controls market pricing!

Spinel is not often used by healers primarily because good specimens are quite costly. However small spinel crystals in a rock matrix, such as pictured here, are affordable, take a polish very well and are durable and attractive for spiritual use. The predominant use for spinel is for heightened clarity and awareness during meditation. It is said that spinel draws energy from the Earth and directs it to the healer for use in spiritual endeavors. Because of this great potential for energy transfer, the suggestions for physical uses are nearly endless. A spinel will dutifully adapt to the needs of the bearer, making it a valuable and powerful addition to any healing crystal collection.

ASTROLOGICAL: *Capricorn,* **CHAKRA:** *Root,* **ELEMENT:** *Fire*

BLUE SPINEL

STAR MICA (Golden Muscovite)

FAMILY: Phyllosilicates

CHEMISTRY: $KAl_2(AlSi_3O_{10})(F,OH)_2\ SiO_2$

HARDNESS: 2 Perpendicular &
4 Parallel to Crystal Grain

STREAK: White

SG: 2.75–2.73

STRUCTURE: Monoclinic Crystal

STAR MICA

GOLDEN STAR MICA (GOLDEN MUSCOVITE) is a bladed, crystalline form of muscovite. It shares identical chemistry but forms as sprays of thin, bladed crystals that usually grow as five-pointed stars. As with all micas, the crystals can be cleaved into thin transparent sheets. Star mica is extremely delicate, making large clustered specimens extremely difficult to find. Its most common known occurrence is in Brazil and its coloring can range from deep gold to pale yellow. We have samples in our family collection that display striking and attractive masses of crystals, and visitors are drawn to them readily. They are colorful and glowing and have a very unusual visual appeal. All micas are extremely fragile and should be handled infrequently and with great care.

Star Mica is a great stone to have nearby when studying, especially when under pressure in school. It is an excellent tool for finding equilibrium and helps with mental, emotional and physical balance. This is a great stone for children! It also acts as a natural appetite suppressor so it may be useful when fasting. Star mica helps to calm anger and turn nervous energy into productive energy.

ASTROLOGICAL: *Cancer,* **CHAKRA:** *Sacral,* **ELEMENT:** *Air*

Related entries in this book: Fuchsite, Lepidolite, Muscovite

STAUROLITE

FAMILY: Nesosilicates

CHEMISTRY: $Fe_2+_2Al_9Si_4O_{23}(OH)$

HARDNESS: 7.5

STREAK: White

SG: 3.74

STRUCTURE: Monoclinic

STAUROLITE

STAUROLUTE is a striking mineral who's crystal prisms almost always form in a distinctive cross-shape. The intersecting, twinned crystals form in metamorphic rocks and in schists, and can accompany such semi-precious crystals as garnet and kyanite. Nice specimens up to palm-size are readily found from mineral dealers; larger ones less common. Staurolite crystals tend to be between 1" to 3" in size and will always be embedded partly in it matrix. Removing the crystals from the matrix will invariably damage the crystals, or destroy the unique cross shape.

Staurolute is the official State Mineral of Georgia, USA. It is also found in Fairy Stone State Park in Virginia, USA. Both "fairy stone" and "fairy cross" are common names for staurolite crystals, however "fairy stone" should not be confused with the more common calcite formation that goes by the same name.

Because of its availability, and associations with the obvious religious symbolism of a cross, staurolite is a popular healing crystal. It is used extensively by both crystal healers and believers in many aspects of spiritualism. It is suggested for forming strong connections to deity, finding compassion, for healing, and for gaining enlightenment. Staurolite is a popular companion for all types of meditation and prayer.

ASTROLOGICAL: *Libra,* **CHAKRA:** *Heart,* **ELEMENT:** *Water*

STICHTITE

FAMILY: Phyllosilicates

CHEMISTRY: $Mg_6Cr_2CO_3(OH)_{16}4H_2O$

HARDNESS: 3.5-4

STREAK: White

SG: 2.20

STRUCTURE: Trigonal hexagonal Crystal

STICHTITE

STICHTITE is a chromium and magnesium carbonate mineral. Its typical color ranges from pink to lilac to a beautiful, deep purple. It is an alteration product that results from weathering and oxidation of serpentinite. Often, specimens are mixed with inclusions of green serpentinite mixed with the purple stichtite. It is not a very common stone by any means and small samples are typically all that can be found in the mineral market. The healing market has yet to find extensive supply, and hence its use is very limited and it remains fairly costly. We have been lucky to occasionally find excellent small specimens and have used some for jewelry purposes, but also retained enough to satisfy our great love of this stone. The purple coloring is vivid and wonderfully textured.

Stichtite is an all encompassing, protective stone which shields the bearer from bad energy and negative influence from groups or individuals. It can be worn, held or pocketed for best effect.

ASTROLOGICAL: *Pisces,* **CHAKRA:** *Crown,* **ELEMENT:** *Air*

Related entry in this book: Serpentinite

TUMBLED STICHTITE

STILBITE

FAMILY: Tectosilicate Zeolites

CHEMISTRY: $NaCa_4(Si_{27}Al_9)O_{72}28(H_2O)$

HARDNESS: 3.5–4

STREAK: White

SG: 2.12–2.22

STRUCTURE: Monoclinic Crystal

STILBITE

STILBITE, like all zeolites, has a high content of water. It usually forms from hydrothermal activity in cavities of basalt or in other igneous rocks. It is often closely associated with apophyllite and scolecite. Stilbite occurs in a multitude of pale colors including yellow, brown, red or orange, but is most commonly dull white with a slight pinkish tint and a pearly shine. In fact, its name is derived from the Ancient Greek word, "stilbe" which translates to "luster" or "sheen". Deposits of stilbite have been found in Australia but most occurrences are found throughout the Deccan Traps region of India. Stilbite and other zeolites are commercially used for filtration of gasses, water and other liquids. Zeolites have a permeable structure and consistent pore size which allows the passage of specific fluids or gasses while filtering-out undesirable contaminants. In India, increased mining activity for collector's grade specimens has brought an abundance of beautiful zeolites to the market. Once rare, these minerals are now easy to find and fairly inexpensive. The varied and striking crystal forms make impressive display items.

Stilbite is useful as a preparation tool and focus aid for Reiki healers. It clears the mind and prepares one for productive and deep meditation. It opens communication with the subconscious and promotes healthy dream states. Stilbite awakens the desire and inspiration for creative and artistic expression. Stilbite is used by physical healers to treat brain disorders and skin problems.

ASTROLOGICAL: *All,* **CHAKRA:** *Third Eye,* **ELEMENT:** *Air*

Related entries in this book: Cavansite, Scolecite

SUGILITE

FAMILY: Cyclosilicates

CHEMISTRY: $KNa_2Fe_3 + 2(Li_3Si_{12})O_{30}$

HARDNESS: 5.5

STREAK: White

SG: 2.75

STRUCTURE: Hexagonal

SUGILITE

SUGILITE is *(along with purpurite and stichtite)*, another of our favorite, beautiful, purple minerals in the family collection. It was named in 1976, in honor of Kenichi Sugi, who first discovered the mineral in Japan. Upon discovery, very little effort was made to mine, market, or use the mineral, but in subsequent years large deposits of sugilite have been found in South Africa, and some quantities are finding their way around the world and into the hands of more people. While still not very common in mineral collections, sugilite has become most popular with crystal healers. Perhaps this is because quality specimens available are mostly small, and display-worthy pieces are scarce and costly. Even so, small, tumbled sugilite stones are not too expensive. Small specimens are typically not polished yet they display deep and rich purple hues.

In the crystal healing realm, sugilite is considered to be an exceptional, all-around nurturing crystal. Physically…spiritually…emotionally, it can be carried, held, or displayed for its purposes. Wearing sugilite as a talisman is a symbol of one manifesting their personal power or regaining inner-strength. Sugilite has been suggested to be used to manifest health of the lungs, airways, and sinuses.

ASTROLOGICAL: *Libra*, **CHAKRA:** *Crown*, **ELEMENT:** *Air*

SULFUR

FAMILY: Native Element, Atomic #16

CHEMISTRY: S_8

HARDNESS: 2-2.5

STREAK: Yellow

SG: 2.07-2.15

STRUCTURE: Orthorhombic Crystal

SULFUR

SULFUR occurs abundantly in nature as a pure element, but more often as a component in sulfide or sulfate minerals. Its crystals make a beautiful addition to any collection due to their deep and rich yellow color and bright sheen. Sulfur forms in deposits near volcanic activity and often near hydrothermal vents. Sulfur is mostly mined today in Sicily, Chile and Poland and in Ohio and Michigan in the United States. In ancient times sulfur was widely known around the world and is mentioned as "brimstone" in some biblical translations. Sulfur was a necessary ingredient in the early production of black gunpowder and still has many practical uses commercially and medicinally. Sulfur is extremely soft and sensitive to moisture, so great care should be taken when displaying or handling it. Most mineral shops or crystal retailers do not sell sulfur because of its fragile nature, but we have found beautiful and inexpensive specimens from scientific and educational supply sources.

Sulfur is used to purify and detoxify environments, focus one's thoughts, and cleanse energy fields. It can be helpful for purification before spiritual rituals as well. Sulphur is used as a shield against hexes and curses, and to free one from negative power or unwanted influence. It treats skin problems like eczema, acne and psoriasis. It can improve digestion and provide relief from arthritis, backaches and sore feet. Sulfur relieves side-effects from chemotherapy, infections and colds. Sulfur has been recommended for detoxifying the body from harmful substances or toxins.

ASTROLOGICAL: *Libra*, **CHAKRA:** *Solar Plexus*, **ELEMENT:** *Fire*

SUNSTONE

FAMILY: Tectosilicates

CHEMISTRY: $(Ca,Na)[(Al,Si)_2Si_2O_8]$

HARDNESS: 6.5–7

STREAK: White

SG: 2.07–2.15

STRUCTURE: Triclinic Crystal

GOLDEN SHEEN SUNSTONE

SUNSTONE is a plagioclase feldspar silicate mineral, once found only in a single location in Norway. Sunstone has become quite popular with collectors and healers in recent years due to new availability from deposits in Oregon, and North Carolina in the United States, and also from Russia. The beautiful, mixed reddish-orange hues of sunstone come from varying concentrations of copper. A good quality piece of sunstone will be somewhat translucent and display a strong inner sheen and flashes of light reflecting off of the minute copper inclusions. Sunstone is sometimes referred to as aventurine sunstone, but this is a misnomer, since true aventurine is a quartz silicate mineral, not a feldspar.

Sunstone brings luck and good fortune to the bearer. It is a cheerful stone that bring feelings of joy as it cleanses and energizes the chakras. It opens the mind and focuses intuition, allowing clearer vision and awareness. It dispels fear, alleviates stress and increases energy. Sunstone is a powerful aid to Reiki healers and spiritual guides, used to regenerate personal energy after sessions. Physically, it treats sore throats and laryngitis and reduces stomach pains and ulcers. Sunstone is also used by healers to treat deteriorating bone cartilage, weakness of bones, and osteoporosis.

ASTROLOGICAL: *Leo,* **CHAKRA:** *Sacral,* **ELEMENT:** *Fire*

Related entries in this book: Feldspar, Labradorite, Moonstone

SUPER 7
(Amethyst, Clear Quartz, Smoky Quartz, Cacoxenite, Rutile, Goethite and Lepidocrosite)

FAMILY: Mixed Mineral

SUPER 7

SUPER 7 is a naturally occurring crystalline silicate variety of quartz. It is a Super seven is a descriptive name for a mixed-mineral found in particular amethyst beds. The seven minerals are: amethyst, goethite, cacoxenite, rutile, clear quartz, lepidocrocite, rutile, and smoky quartz. First identified in a single deposit in Brazil, super seven is now been identified in Madagascar and Australia. The proportions of each mineral in super seven vary greatly, and some specimens may not actually include all seven. It is also difficult to visually distinguish the individual minerals due to similarity of features, opacity of the specimen and inclusions of other minerals. Super seven is sold in much the same forms as amethyst specimens; slabs, masses, tumbled stones, jewelry, etc., and it can be readily found at mineral dealers as well as from crystal healing sources.

As a healing crystal, super seven is very popular and considered to be an energetic and powerful healer. Used primarily for raising physical or spiritual energy levels, it is also a stone for harmony and peace. Being a harmonious crystal it can be used for finding balance, forging stronger relationships with others, and maintaining focus and clarity of thought.

ASTROLOGICAL: *All*, **CHAKRA:** *All*, **ELEMENT:** *None*

Related entries in this book: Amethyst, Quartz, Smoky Quartz

TANZANITE

FAMILY: Sorosilicates

CHEMISTRY: $(Ca_2Al_3(SiO_4)(Si_2O_7)O(OH)$

HARDNESS: 6.5

STREAK: White

SG: 3.10–3.38

STRUCTURE: Orthorhombic Crystal

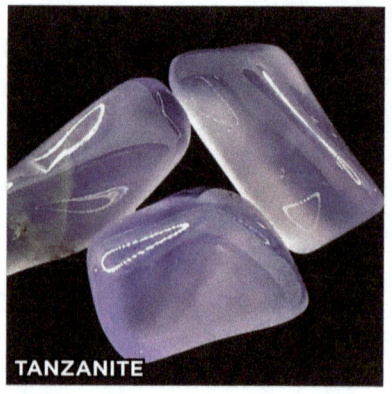

TANZANITE

TANZANITE is a rare, bluish-purple variety of zoisite *(calcium aluminum hydroxyl silicate)*. It is found only in the area surrounding Mt. Kilimanjaro in Tanzania, and was discovered there in 1967 by gold prospectors. It displays a rare optical effect called trichroism, which means that depending on the viewing angle and lighting conditions, tanzanite can appear any one of three colors, deep blue, purple or red. Its natural color is actually reddish-brown and needs to be treated at very high temperature to bring out blue and purple hues. Higher heat-treating can produce finer, richer, gem quality stones, but they will display only dichroism, changing between purple and blue. These specific colors vary by the degree of metallic content in the stone, usually aluminum and sometimes chromium. Tanzanite is a popular jewelry stone but care must be taken when shaping and polishing because it is brittle and chips easily. It is also sensitive to sudden temperature change, which can cause tanzanite to fracture. Because of its scarcity and its commercial value for jewelry, it is difficult to find specimens suitable for collectors or healers. Most examples found in the general retail market are either quite small, or inferior in color intensity. Recently strings of small beads and tanzanite stone chips have made their way to the market from sources in China.

Tanzanite is an excellent crystal for those who are new to crystal healing or to those beginning their exploration of psychic powers. It is a very gentle and protective crystal, and allows one's experiences to develop slowly and in proper progression. Physically, tanzanite is used to promote the repair of damaged cells, especially in the skin and hair, and it protects against the side-effects of surgery or chemotherapy/radiation.

ASTROLOGICAL: *Pisces,* **CHAKRA:** *Throat,* **ELEMENT:** *Water*

TEKTITES

Tektites present the collector with a unique opportunity to explore the interaction of Earthly and unearthly objects upon each other. Tektites were formed from collision of celestial materials *(meteoroids)* with the surface of our planet, thus, the other name for tektite is "impactite".Because of the extreme heat-energy caused by a high-speed, meteorite impact, there is a great deal of melt of both the impactor and the surface rock. The subsequent ejection of molten silicate material from the impact site cools quickly as it 'splashes" away from the impact. Molten silicate rock that cools too quickly to form crystals of interlocking molecules, is said to be amorphous *(without form)*, and is technically glass. All tektites can be categorized as natural glass. Due to trace elements present in the impactor and at the impact site, tektites take-on many forms and display many colors. Properties they all share are brittleness, hardness the same as any glass *(approx. 5.5 on Moh's scale)*, and will be translucent if thin enough sections are examined. Tektites will also be of non-uniform shape and come in a vast range of sizes.

Related entries in this book: Lechitelierite (Fulgarite), Obsidian (both natural and manufactured)

SEE PAGE #S BELOW FOR INDIVIDUAL TEKTITES

TIBETAN TEKTITES

FAMILY: Mineraloid

CHEMISTRY: Variable, Mostly Silicate

HARDNESS: 5-6

STREAK: White

SG: 2.28-2.51

STRUCTURE: Amorphous Rock

TIBETAN TEKTITES

TIBETAN TEKTITES OR CHINESE TEKTITES are rare, natural glass objects composed primarily of silica, and are formed due to comet, meteor or asteroid impact. The intense heat and energy from impact, melts rock from both the impactor and the surface of the Earth and ejects molten material into the atmosphere. As the molten ejecta falls back to Earth, it cools too rapidly to form regular crystalline structures. Thus, tektites cannot be properly categorized as a crystalline rock but rather as a glass mineraloid. Tektites are found in a wide variety of shapes and sizes, some of the most desirable forms being teardrop, dumbbell and bullet shaped rocks. Other tektites are "flying saucer" shaped, indicating little or no spin as they cooled in the atmosphere. Dumbbell shaped tektites were probably spinning rapidly enough to stretch and would eventually have split apart into bullet shapes, given more time. Tektites have been found on every continent except South America and Antarctica. Commonly, tektites on the market are sold as "Tibetan tektites" but there is no way to confirm their origin, as all tektites have identical chemistry and similar appearance. Unfortunately, the moniker "Tibetan" may be a marketing ploy to imply a more spiritual product. Wherever tektites have been discovered, they are found in concentrated group-ings, similar in many ways to meteorite strewn patterns. It is assumed that these groupings are the result of an impact event. Because of their unique origin and their connection to extra-terrestrial matter, tektites are an unusual and welcome addition to any collection. We regard our tektite specimens as very special and with a sense of awe.

Tektites raise one's vibration to higher levels and strengthen the aura, making it a good tool for personal transformation and elevation of the spirit. Tektites can be useful during meditation to expand consciousness. Tektites accelerate healing from illness and can help reduce fever, chills and body aches.

ASTROLOGICAL: *Aries,* **CHAKRA:** *Third Eye,* **ELEMENT:** *Air & Fire*

THULITE (Zoisite Variety)

FAMILY: Sorosilicates

CHEMISTRY: $(Ca_2)(Al,Mn_3+3)(Si_2O_7)$ $(SiO_4)O(OH)$

HARDNESS: 6.0–7.0

STREAK: White

SG: 3.15

STRUCTURE: Orthorhombic

THULITE

THULITE (ZOISITE VARIETY) is wonderfully hot-pink, opaque mineral, part of the broad family of zoisite minerals. This family includes such popular and beautiful varieties such as purple tanzanite, and green zoisite *(popularly collected as ruby-in-zoisite crystals)*. Thulite is soft and is rarely shaped or polished, it large and small raw specimens are available for display. Veins of white calcite are common inclusions in larger specimens. Some jewelry pieces are made of pink thulite because it a very attractive stone, but care should be taken as it tends to fracture and chip easily.

Thulite is mined primarily in Norway where it was named for the mythical island of Thule, which was indicated on ancient maps as being to the North of all known lands. Thulite it is also found in North Carolina, USA., however these specimens are usually not as vivid in color as the Norwegian stones. Thulite makes an excellent and vibrant addition to any collection.

Healers use thulite for stimulating passion and increased intimacy. The uses for personal relationships are many, from forging business relationships to improving the ability to speak in public or feel at ease communicating inner-feelings.

ASTROLOGICAL: *Gemini,* **CHAKRA:** *Third Eye & Throat,* **ELEMENT:** *Earth*

Related entries in this book: Tanzanite, Zoisite

TIGER'S EYE

FAMILY: Silicates

CHEMISTRY: Variable, Mostly Silicate (SiO_2)

HARDNESS: 7

STREAK: White

SG: 2.64–2.71

STRUCTURE: Trigonal Hexagon

RAW TIGER'S EYE SLAB

TIGER'S EYE is a composite stone composed of alternating layers of iron-stone or hematite, and colorful crocoite filaments *(natural asbestos)* that have been pseudomorphed *(replaced)* by quartz. Because of the layered arrangement and properties of each, they shine with a deep, characteristic chatoyance *(cat's eye effect)*. Many specimens show exceptionally brilliant flashes of light. It is a common stone and very affordable, making it a favorite for jewelry and for collecting. Gold is the predominant color variety of tiger's eye found naturally, blue being second although very rare. Rarest of all is natural red tiger's eye, however due to the ease of heat-treating and the low temperature required to turn gold tiger's eye to red, it is extremely popular and widely available commercially. We had great success treating both tumbled and raw, gold tiger's eye in our kitchen oven, turning them beautifully red. Instructions for the process follow. Please take care when dealing with heat and ovens. Any heated stone may shatter or explode due to inner cavities or fractures, so please follow our advice below about wrapping the stones, and keeping them wrapped until they cool. Always supervise children.

How to make red tiger's eye stones: Wrap some gold tiger's eye stones in several layers of aluminum foil and place them on a tray, in an oven at 250°F. Increase the heat 50°F every half-hour until the oven reaches 550°F. Continue heating at 550°F for 2 more hours. Then turn the oven off and let the stones cool slowly for 2 hours untouched so they do not shatter from cooling too quickly. The process can be repeated for a longer period of time if the stones have not turned color, but higher temperature will not have any greater effect. Good Luck!

GOLD TIGER'S EYE

RED TIGER'S EYE

GOLD TIGER'S EYE offers faithful protection to the bearer and brings good luck. It stabilizes mood swings and releases tension. Gold tiger's eye treats the eyes, throat and reproductive organs, alleviates pain and speeds the healing of broken bones and strengthens the back. It is a favorite lucky stone for children and for adults who appreciate bright, shiny objects *(this includes the author)*.

ASTROLOGICAL: *Capricorn,* **CHAKRA:** *Sacral,* **ELEMENT:** *Fire*

RED TIGER'S EYE increases personal willpower, inner-strength and resolve. It brings soothing calm to the soul and releases stress and anxiety. It promotes patience, and helps those who tend to make rash decisions. Red tiger's eye cools over-active romantic desires.

ASTROLOGICAL: *Capricorn,* **CHAKRA:** *Base,* **ELEMENT:** *Fire*

BLUE TIGER'S EYE (HAWK'S EYE) is said to share the spiritual properties of both gold and red tiger's eye, with the exception that is also suggested for stimulating under-active romantic desires.

ASTROLOGICAL: *Capricorn,* **CHAKRA:** *Throat,* **ELEMENT:** *Fire*

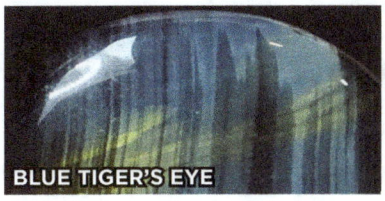

BLUE TIGER'S EYE

TIGER IRON

FAMILY: Oxides & Cryptocrystalline Silicate

CHEMISTRY: Mixed Mineral (Fe & SiO_2)

HARDNESS: 6.5–7

STREAK: White

SG: 2.60–2.71

STRUCTURE: Mixed Crystal

TIGER IRON

TIGER IRON is a composite rock, formed mainly from gold tiger's eye, red jasper, and black hematite. Although it is a composite, it is sometimes used as a semi-precious gemstone because of its rich coloring and distinctive banding. Tiger iron is most commonly mined and exported to the collector's market from Australia, Brazil, Mexico and from Minnesota and Michigan in the United States. Polished specimens bring out the bright banding of the tiger's eye mineral which is itself, a quartz replacement of crocoite *(natural asbestos)*. Every piece of tiger iron is different; some show a predominance of red jasper, and others are more inclusive of tiger's eye or hematite. It is a popular and inexpensive material and easy to find. Tiger iron is extremely common in areas of Michigan's Upper Peninsula mining region in the USA, however specimens with inclusions of tiger's eye are relatively rare there. Commercially and geologically, hematite with alternating bands of red jasper is referred to as a Banded Iron Deposits *(BIF)* and is the major source of iron ore worldwide. BIF's are extremely ancient and indicative of the time in Earth's ancient past when oceans were iron-rich and the atmosphere first gained oxygen from outgassing of the planet. Oxygen and iron reacted, and the heavy, iron-oxide molecules precipitated to the ocean floor. Later, uplift as well as modern mining, has exposed these valuable iron deposits.

Tiger iron is a great stone to use for encouraging creative and artistic abilities. It is helpful when recovering from exhaustion, especially emotional or mental burn-out or family stress. Tiger iron is used to boost the body's energy and help with assimilation of B vitamins, and to balance the red to white blood cell count. It strengthens muscles, and aids in producing the natural hormones, endorphins and enzymes needed for a healthy physical being.

ASTROLOGICAL: *None,* **CHAKRA:** *All,* **ELEMENT:** *Earth*

Related entries in this book: Hematite, Red Jasper, Tiger's Eye

TOPAZ

FAMILY: Nesosilicates

CHEMISTRY: $Al_2SiO_4(F,OH)_2$

HARDNESS: 8.0 (Defining)

STREAK: White

SG: 3.49–3.57

STRUCTURE: Orthorhombic Crystal

IMPERIAL TOPAZ IN MATRIX

TOPAZ is an extremely hard, transparent and attractive aluminum silicate gemstone. Pure topaz is colorless, but precious examples exist in a wide range of beautiful hues, and are tinted by mostly metallic chemical impurities. Like so many precious gemstones, topaz is found in pegmatite intrusions of granite, or in cavities in rhyolitic igneous rock flows. Topaz can be found almost worldwide in locations including Mexico, Sri Lanka, Germany and Russia. It is the official state mineral of Utah in the United States. It has always been considered a rare and precious gemstone especially golden and yellow varieties. Ancient Roman history refers to the island "Topazos" *(now St. John's Island)* in the Red Sea, as the source of precious topaz. However it is now believed the word "topaz" referred to any golden-yellow crystal used for ornament at that time. Flawed or non-gem quality topaz specimens can be fairly affordable and are still attractive enough for the casual collector. Even finer examples, still in their base matrix can be purchased for a reasonable price and make most excellent display samples.

Topaz stimulates and realigns the meridians of the body, directing healing energy where it is needed. It promotes an attitude of truth and forgiveness. Topaz is a stone of great joy and generosity, abundance and health, love and good fortune. It promotes honesty, self-realization and self-control, aids with creative problem-solving, and assists in expressing difficult thoughts. For physical healing, topaz is recommended for improving digestion and treating eating disorders, such as anorexia and bulimia. It is beneficial to the nervous system and helps to balance internal organ function and the metabolism.

ASTROLOGICAL: *Leo*, **CHAKRA:** *Sacral*, **ELEMENT:** *Air*

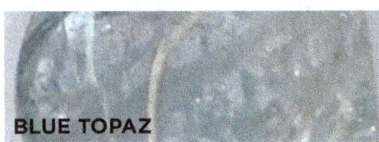

BLUE TOPAZ

TOURMALINE

FAMILY: Cyclosilicates

CHEMISTRY: (Ca,K,Na)
(Al,Fe,Li,Mg,Mn)$_3$(Al,Cr, Fe,V)$_6$–
(BO3)$_3$(Si,Al,B)$_6$O$_{18}$(OH,F)$_4$

HARDNESS: 7–7.5

STREAK: White

SG: 3.00–3.33

STRUCTURE: Trigonal Hexagon Crystal

TOURMALINE VARIETIES

TOURMALINE is a boro-silicate mineral, with many elements included in its extremely complicated chemistry. The high metallic content gives tourmaline its characteristic hardness and relative heaviness. These elements, in varying ratios, are responsible for the many colors for which tourmaline is known. Available as blue, pink, black, purple, green, yellow and even brown crystals, it is classified as a semi-precious stone and is commonly used for fine jewelry. In its raw state, tourmalines have attractive crystal habits, general translucence, and a semi-metallic shine making them extremely popular with collectors.

Tourmaline typically forms in granite, pegmatite intrusions. Elba, Italy has historically been a major source of high-quality and desirable tourmaline colors, occurring in deep shades of blues, greens and pink. These particular tourmalines are even referred to as the elbaite series, and have been found to be rich in alkali lithium content. However, the vast majority of tourmaline, perhaps as much as 95%, is black, brittle and not suitable for jewelry purposes. Black tourmaline is referred to as schorl.

BLACK TOURMALINE

PINK TOURMALINE

BLACK TOURMALINE (SCHORL) is a strong defense against negativity. It is excellent for deflecting harmful energy. It enhances well-being by providing an increase in physical energy, emotional stability, and intellectual acuity. Black tourmaline treats paranoia, dyslexia, and clumsiness. It releases muscle tension and is helpful for spinal health. Since black tourmaline absorbs negative energy eagerly, it should be cleansed in sunlight or moonlight regularly.

ASTROLOGICAL: *Taurus,* **CHAKRA:** *Base,* **ELEMENT:** *Earth*

PINK (AND WATERMELON) TOURMALINE helps release stress and anxiety, and is uplifting during bouts of depressions. Used in conjunction with black tourmaline, it stops obsessive behavior. It is useful for spiritually sensitive indigo children for its calming qualities, and as a reminder to pause and reflect before taking action. Physically, pink tourmaline reduces travel and motion sickness. It may also assist in restoring damaged nails and hair.

ASTROLOGICAL: *Taurus,* **CHAKRA:** *Crown,* **ELEMENT:** *Fire*

WATERMELON TOURMALINE

GREEN TOURMALINE

TURQUOISE

FAMILY: Phosphates

CHEMISTRY: $CuAl_6(PO_4)_4(OH)_8 4H_2O$

HARDNESS: 5–7

STREAK: Blueish White

SG: 2.59–2.90

STRUCTURE: Triclinic Crystal

BLUE TURQUOISE

TURQUOISE is a fairly hard, blue-green, phosphate mineral that gets its famous color from content of copper and aluminum. It is most often found in arid regions and is primarily an oxide of other minerals. Higher iron content and exposure to more moisture is probably responsible for adding more green hues to some turquoise. It usually forms in fissures in feldspar rock or as small nodules in aluminum-rich, igneous intrusions near the surface. It is a fairly rare mineral, yet extremely popular and highly valued for many centuries as a gemstone for jewelry. Archaeological evidence suggests that turquoise has been used as a decorative stone for at least 8000 years. The name turquoise comes from an old French word for "Turkish", because the first turquoise brought to Europe originated in Turkey. Ancient mining sites in parts of Persia *(Iran)* and the Sinai Peninsula *(Egypt)* still exist. The most productive locations for mining turquoise are currently in the Southwestern United States, in Colorado and New Mexico.

Turquoise is a stone of purification for cleansing the spirit and body. It dispels negative energy and protects against negative influence. Turquoise protects against pollution and contaminants in air and water. It balances the body's energy and aligns the chakras, stabilizes mood swings and instills meditative tranquility. A booster for the immune system, turquoise fights illness and acts as an anti-inflammatory and detoxifier. It is also suggested for relieving abdominal cramps and pains.

ASTROLOGICAL: *Scorpio,* **CHAKRA:** *Heart,* **ELEMENT:** *Air & Fire*

GREEN TURQUOISE

UNAKITE

FAMILY: Tecto/Cryptocrystaline Silicate

CHEMISTRY: $KAlSi_3O_8$ & SiO_2

HARDNESS: 6.5–7

STREAK: White

SG: 2.60–2.70

STRUCTURE: Mixed Crystal

UNAKITE

UNAKITE is a mixed-mineral rock composed of pink, orthoclase feldspar, green epidote, and white or clear quartz. It occurs in a variety of shades and is usually mottled or speckled in appearance. While it can be found in several locations in China, Brazil and South Africa, it is found most abundantly in the United States. First discovered in the 1800's in the Unakis Mountains in North Carolina, it can sometimes be found laying loose in alluvial deposits in Virginia. Unakite has become an extremely popular semi-precious gemstone, partly because it is hard, durable and very colorful. It takes well to polishing and makes very attractive carvings and figures. Some examples of unakite on the market lack the characteristic, pink inclusions of feldspar and can be easily mistaken for green epidosite. We have several unakite specimens in our collection, and those that appear to lack any visible pink inclusions can easily be identified with a magnifying lens that clearly shows that small grains of feldspar are present.

Unakite is a stone of higher vision. It is spiritually uplifting but provides grounding when needed. It facilitates the release of emotions that might be blocking spiritual growth. Unakite is a family favorite for relief during allergy season! It also treats reproductive disorders and nurtures the health and well-being of new mothers and their babies. Good for healthy hair and for re-growing skin after scrapes or injury.

ASTROLOGICAL: *Scorpio*, **CHAKRA:** *Heart*, **ELEMENT:** *Fire*

Related entries in this book: Epidote, Feldspar (orthoclase)

VANADANITE

FAMILY: Phosphates

CHEMISTRY: $Pb_5(VO_4)_3Cl$

HARDNESS: 3

STREAK: Brownish Yellow

SG: 6.61–7.32

STRUCTURE: Hexagonal Crystal

VANADANITE

VANADANITE is a member of the apatite group of phosphates, and is composed primarily of lead and vanadium. It is quite dense and forms beautiful, orange-red crystals. It has a slight translucence, which is very uncommon for such a heavy mineral. It occurs naturally in the Southwest United States and in Mexico, usually in barite or calcite deposits. Vanadanite is a minor source of lead, but it is mined for its vanadium content which is used as an additive to steel for making durable hand and power tools. Because of its high lead content, vanadanite should not be handled frequently or by children. **Please wash your hands after touching.**

Vanadanite is a stone of persistence and creativity. Excellent for getting through writer's block and creative impasses. When energy is dwindling, vanadanite is used as a booster that helps maintain focus on the task at hand. Specimens with the strongest red color are said to indicate love and romance. Suggested as a safe-guard against radiation and the ill-effects of chemotherapies. It also provides defense against the aches and pains of aging.

ASTROLOGICAL: *Virgo*, **CHAKRA:** *Third Eye*, **ELEMENT:** *Fire*

VESUVIANITE (Idocrase)

FAMILY: Sorosilicates

CHEMISTRY:
$Ca_{10}(Mg, Fe)_2Al_4(SiO_4)_5(Si_2O7)_2(OH,F)_4$

HARDNESS: 6.5

STREAK: White

SG: 3.32

STRUCTURE: Tetragonal

VESUVIANITE

VESUVIANITE (IDOCRASE) draws its name from Mount Vesuvius, the famous volcano in Naples, Italy where it was first identified in metamorphic rocks near contact zones with igneous intrusions. Vesuvianite can form fine crystals but is more commonly found as an opaque rock with varied greenish hues. It is fairly hard and takes to shaping and polishing fairly well, yet rarely achieves a high-luster. It may be more commonly sold as small tumbled samples than as collector-sized specimens. Vesuvianite is found Italy, Kenya and in California, USA. Some quantity of gem-quality crystals now coming primarily from Kenya and Tanzania are typically shaped into cabochons or beads rather than faceted due to rather low translucence and only a semi-greasy luster.

For crystal healing, vesuvianite heals a heavy heart and helps one return to opportunities for true love. Vesuvianite removes feelings of fear and trepidation, and helps overcome blockages and the hesitations that come from self-doubt.

ASTROLOGICAL: *Sagittarius,* **CHAKRA:** *Heart,* **ELEMENT:** *Fire*

WAVELLITE

FAMILY: Phosphates

CHEMISTRY: $Al_3(PO_4)_2(OH,F)_3 \cdot 5H_2O$

HARDNESS: 3.5

STREAK: White

SG: 2.36

STRUCTURE: Orthorhombic

WAVELLITE

WAVELLITE was first described in England in the early 19th Century but it's most notable location is currently in the Mount Ida area of Arkansas, USA. It is a remarkable mineral in that wavellite crystals form mainly as small spheres in a matrix of aluminum-rich metamorphic rock. Display pieces are abundant and can display both intact spheres dotting the surface of the specimen, and also where the spheres have been cut-through or broken, they display circles with striking radial patterns. Wavellite crystals range in size from 1/16" to 1/2" and in color from lime to forest green. The combination of form and color make wavellite a truly unique piece for the mineral collection!

Wavellite is not very often used in crystal healing but has been suggested for decision-making and improving focus. Rejuvenation and increased energy are also possible uses for the physical body.

ASTROLOGICAL: *Aquarius,* **CHAKRA:** *Hearty,* **ELEMENT:** *Water*

ZEBRA STONE

FAMILY: Cryptocrystalline Silicates

CHEMISTRY: SiO_2

HARDNESS: 6.5–7

STREAK: White

SG: 2.51–2.80

STRUCTURE: Trigonal Hexagon Crystal

ZEBRA STONE

ZEBRA STONE has distinctive stripes that were created by deposits of ancient bacterial colonies that filled ripple marks in a river bed or in shallow water. The alternating pattern of sediment and biological debris eventually hardened into rock, due to the chemical reaction with silica from the water source. This actually makes it a form of opal chalcedony, and is therefore related to other silicate minerals such as turritella agate and butterstone, wherein silicate has replaced biological matter. Zebra stone should not be confused with zebra jasper which has a different white with black, or white with brown, striped pattern. Zebra stone is in fairly limited supply at mineral and collector's shops since it is mined in only a single location in Australia. We have seen it sold in raw or top-surface polished forms, in a variety of sizes and shapes. Each stone is uniquely patterned, and the color can vary considerably. It is a very interesting and visually appealing stone for display.

Zebra stone can be used to connect to the Earth and to the energy and love of the universe. By making these connections, it becomes easier to discover one's true self and realize inner potential. Zebra stone is effective at protecting and strengthening the aura. Physically, zebra stone increases stamina, endurance and perseverance. It can be useful for treating bone injuries, osteoporosis, teeth and gums. Zebra stone relieves muscle twitches and spasms, and since it provides relief for all of the body's muscles, it can ease mild heart palpitations.

ASTROLOGICAL: *All,* **CHAKRA:** *Third Eye,* **ELEMENT:** *Air*

ZIRCON

FAMILY: Nesosilicates

CHEMISTRY: $Zr(SiO_4)$

HARDNESS: #

STREAK: 7.5

SG: 4.6

STRUCTURE: Tetragonal

BLUE ZIRCON

ZIRCON is a very hard, lustrous, gem-quality crystal that happens to be much overlooked or mis-identified. Since the marketing in the 1970's of synthetic diamond substitute named "Cubic Zirconia", the public has generally shrugged-off true zircon as being an inferior or "fake" gemstone. Nothing could be farther from the truth! Natural zircon ranges in shades of deep blue, to vivid red, to incredibly clear crystals, free from fractures or inclusions. It is an ideal faceted gemstone used for rings, brooches, pendants, etc. and is more affordable than gems of similar color, such as topaz or sapphire. Raw crystals attached to their rock matrix make wonderful collector specimens. Natural zircon crystals can be found in all types of rock; metamorphic, igneous, and sedimentary, and range in size from microscopic to 3 or 4 centimeters. Zircon is extremely durable and has survived the most extreme geologic conditions for eons. Zircon can contain incredibly small traces of radioactive uranium and thorium *(totally safe to humans)*, which are however enough to accurately date the crystals. In the late 20th Century, zircon crystals found in Australia were dated to an origin over 4.4 billion years ago, making them the oldest known minerals in our planet.

Zircon has been used for centuries as a healing crystal for a wide variety of physical ailments of the internal organs, especially the liver and lymph systems. It is a purification stone and as such helps the bearer with clarity, energy, honesty and integrity.

ASTROLOGICAL: *Taurus*, **CHAKRA:** *All*, **ELEMENT:** *Air*

HYACINTH TOPAZ

INDICES

HARDNESS INDEX (MOHS SCALE)

The Mohs scale of mineral hardness is based on the ability of one mineral sample to scratch another. The scale was created in 1812 by the German geologist and mineralogist Friedrich Mohs. The scale is not linear, in that the defining minerals for each hardness number were based on availability to Herr Mohs and are common enough for most mineralogists to acquire for testing purposes. The scale is not based on equal values of hardness between each hardness number. Remember hardness does not correspond to tensile *(bending)* strength or fracture *(cracking or cleaving)*. "Hard" minerals can still break or shatter easily, so be careful with your own specimens!

The way to determine hardness is to find the hardest defining material that your sample can scratch. Then find the softest defining material that can scratch your sample. Your specimen will have a hardness in between those two defining numbers.

The scale is based on the following list of defining sample minerals. Most are commonly available:

1-Talc	5-Apatite	9-Corundum
2-Gypsum	6-Orthoclase Feldspar	10-Diamond*
3-Calcite	7-Quartz	
4-Fluorite	8-Topaz	

Diamond is not necessary to include in a test kit, since no other mineral can scratch it, and corundum (9) can scratch all other minerals besides diamond.

There are also some common items that can be used to approximate a specimen's hardness, such as:

Graphite *(pencil "lead")*—hardness 1.5
Fingernail—hardness 2.5
Copper penny—hardness 3.5
Knife blade—hardness 5.1
Hardened steel blade—hardness 5.5
Glass—hardness 5.5
Steel file—hardness 6.5

INDEX OF SPECIMENS BY HARDNESS (MOH'S SCALE)

HARDNESS 6 CON'T:
Labradorite 6-6.5
Moonstone 6
Petalite 6-6.5
Prehnite 6-6.5
Pyrite 6-6.5
Rainforest Rhyolite 6.5-7
Rutile 6.5
Sunstone 6.5-7
Tanzanite 6.5
Unakite 6.5-7
Zebra Stone 6.5-7

HARDNESS 7:
Agates 7.0
Amethyst 7.0
Aquamarine 7.5-8
Aventurine 6-7
Axinite 6-7.5
Carnelian 7
Chalcedony 7
Chrysoprase 7
Citrine 7
Dumortierite 7.5-8
Emerald 7.5-8
Lemurian Crystal 7
Mookaite 7
Onyx 7
Petrified Wood 7
Pietersite 7
Quartz 7.0 (defining)
Shiva Lingam 7
Tiger's Eye 7
Tourmaline 7-7.5

HARDNESS 8:
Aquamarine 7.5-8
Dumortierite 7.5-8
Emerald 7.5-8
Topaz 8.0 (defining)

HARDNESS 9:
Corundum (ruby & sapphire) 9.0
(defining)

STREAK COLOR INFORMATION

Streak plates are one method of identifying minerals. The other common method is hardness testing, using the Mohs scale *(see previous section in this book)*. The streak *(or "powder color")* of a mineral is the color of the powder left by scratching it on a standardized surface. While the visual color of a rock sample varies from piece to piece, the streak color for a particular substance will always be the same. This is why it is a valuable and predictable method of identification. Some minerals seem to leave no streak and are classified as white or colorless.

The standard surface for streak testing that most geologists agree upon is un-glazed, white porcelain. It has an approximate hardness of 7 *(Mohs scale)* which means minerals harder than 7 will not leave a streak. However most harder minerals leave only a white streak anyway, so scratch testing alone is not a sufficient method for identification.

Since silicates compose over 90% of the weight of the earth's crust, most commonly found rocks are composed mainly of this class of minerals, and all will have the same characteristic white or colorless streak. The vast majority of rocks in most collections *(ours included)* will be silicates. For the purposes of this guide, only rocks and minerals with non-white streak test colors are listed.

STREAK COLOR INDEX OF SPECIMENS *(NON-WHITE ONLY)*

MINERALS WITH STREAK COLOR YELLOW:
Astrophyllite, Gold, Sulfur

MINERALS WITH STREAK COLOR GREEN:
Crysacolla *(blue-green)*, Malachite *(light green)*, Pyrite *(green to light brown)*

MINERALS WITH STREAK COLOR BLUE:
Azurite, Turquoise, Shattuckite

MINERALS WITH STREAK COLOR BROWN:
Boji Stone, Pyrite *(green to light brown)*, Jet, Vanadanite *(brownish yellow)*

MINERALS WITH STREAK COLOR BLACK OR GRAY:
Bornite *(grayish black)*, Epidote *(gray-light black)*, Galena *(lead gray)*

MINERALS WITH STREAK COLOR RED OR PINK:
Copper *(rose pink)*, Moqui Marble *(bright to dark red)*, Hematite *(dark to bright red)* Magnetite *(lodestone) (dark to bright red)*

MINERALS WITH STREAK COLOR SILVER:
Bismuth, Silver

SPECIFIC GRAVITY INFORMATION

Specific Gravity *(SG)* is a measurement that describes the density of minerals. It is another valuable diagnostic for identification. Specific gravity quantifies the density of a mineral as compared to the density of water. Water has an SG of 1.0.

Most minerals that are found on the Earth's crust consist primarily of silicates and will have an SG of around 2.72. Therefore these minerals are considered to be of "average" density. An alphabetical listing of minerals with their respective SG follows. There is also a listing of minerals in order of magnitude from lowest SG to highest. This is a handy way to compare any specimen to another. As you familiarize yourself with the property of specific gravity, it is worthwhile practice and fun to handle specimens from your own collection and compare how heavy or light they feel in your hand. You will notice that minerals such as jet or amber feel extremely light; less than half the SG of quartz or amethyst. A sample of kyanite or a rough garnet will weigh approximately one and a half times the weight of quartz. Pyrite or fool's gold is twice the SG of quartz, and copper is fully three times the SG of quartz. Minerals with SG below 2 and above 8 are relatively uncommon. The following descriptions are commonly used when discussing specific gravity.

SG 1 TO 2	VERY LIGHT
SG 2 TO 2.5	LIGHT
SG 2.5 TO 3	AVERAGE
SG 3 TO 3.5	SLIGHTLY ABOVE AVERAGE
SG 3.5 TO 4	ABOVE AVERAGE
SG 4 TO 5	HEAVY
SG 5 TO 7	VERY HEAVY
SG 7 TO 10	EXTREMELY HEAVY

SPECIFIC GRAVITY ALPHABETIC INDEX OF SPECIMENS

Agate (all)	2.58 - 2.64		Fluorite	3.18 - 3.21
Amazonite	2.56 - 2.59		Fool's Gold (pyrite)	4.95 - 5.10
Amethyst	2.64 - 2.66		Fuchsite	2.85 - 2.88
Andalusite	3.17 - 3.19		Fulgarite	2.50 - 2.71
Angelite	2.80 - 2.90		Fuchsite	2.85 - 2.88
Anyhdrite	2.80 - 2.90		Galena	7.22 - 7.61
Apache Tears	2.30 - 2.60		Garnet	3.80 - 4.25
Apatite	3.17 - 3.23		Garnet (green)	3.70 - 4.10
Apophyllite	2.32 - 2.41		Gold	19.30 - 19.38
Aquamarine	2.84 - 2.91		Goldstone	2.50 - 2.80
Aragonite	2.93 - 2.94		Gypsum Rose	2.31 - 2.33
Astrophyllite	3.20 - 3.40		Halite	2.11 - 2.16
Aventurine	2.64 - 2.69		Heliotrope	2.60 - 2.62
Axinite	3.31 - 3.34		Hematite	5.30 - 5.33
Azurite	3.82 - 3.84		Herkimer Diamond	2.71 - 2.73
Barite	4.20 - 4.22		Howlite	2.53 - 2.59
Bismuth	9.68 - 9.74		Hypersthene	3.10 - 3.33
Bloodstone	2.61 - 2.68		Iceland Spar	2.69 - 2.71
Boji Stone	varies		Jade (Jadeite)	3.30 - 3.40
Bornite	5.08 - 5.11		Jade Nephrite	2.90 - 3.02
Bronzite	3.20 - 3.21		Jasper	2.50 - 2.91
Calcite	2.70 - 2.71		Jet	1.30 - 1.34
Carnelian	2.61 - 2.66		Kyanite	3.53 - 3.68
Celestine	3.96 - 3.98		Labradorite	2.68 - 2.72
Celestite	3.96 - 3.98		Lapis Lazuli	2.40 - 2.91
Chalcedony	2.58 - 2.59		Lazurite	2.40 - 2.91
Charoite	2.54 - 2.56		Lemurian Crystal	2.71 - 2.73
Chiastolite	3.17 - 3.19		Lepidolite	2.80 - 2.90
Chrysocolla	2.00 - 2.40		Lodestone	4.90 - 5.20
Chrysoprase	2.60 - 2.64		Magnetite	4.90 - 5.20
Citrine	2.65 - 2.66		Malachite	3.70 - 4.01
Copal	1.03 - 1.08		Mica (star)	2.75 - 3.00
Copper	8.96 - 8.97		Mookaite	2.64 - 2.69
Corundum	3.99 - 4.00		Moonstone	2.54 - 2.59
Creedite	2.69 - 2.71		Moqui Marble	varies
Diopside	3.30 - 3.32		Nuummite	2.92 - 3.03
Dumortierite	3.35 - 3.37		Obsidian	2.40 - 2.60
Emerald	2.75 - 2.78		Onyx	2.58 - 2.64
Epidote	3.30 - 3.50		Opal (green)	1.98 - 2.25
Feldspar	2.55 - 2.63		Peacock Ore	5.08 - 5.11

Peridot	3.20 - 4.30
Petalite	2.42 - 2.51
Petrified Wood	2.58 - 2.91
Pietersite	2.58 - 2.66
Prehnite	2.82 - 2.94
Pyrite	4.95 - 5.10
Quartz	2.71 - 2.73
Rainforest Rhyolite	2.41 - 2.62
Rhodocrosite	3.51 - 3.70
Rhodonite	3.40 - 3.70
Ruby	3.99 - 4.01
Ruby & Fuchsite	varies
Rutile	4.23 - 4.25
Sapphire	3.99 - 4.00
Scolecite	2.16 - 2.40
Selenite	2.30 - 2.33
Septarian	varies
Serpentine	2.50 - 2.65
Shattuckite	3.86 - 4.15
Shiva Lingam	2.70 - 2.76
Silver	10.14 - 11.00
Sodalite	2.27 - 3.33
Stilbite	2.12 - 2.22
Spar (Iceland)	2.69 - 2.71
Sulfur	2.07 - 2.15
Sunstone	2.69 - 2.72
Tanzanite	3.10 - 3.38
Tektites	2.28 - 2.51
Tiger's Eye	2.64 - 2.71
Tiger Iron	2.60 - 2.71
Topaz	3.49 - 3.57
Tourmaline	3.00 - 3.33
Turquoise	2.59 - 2.90
Unakite	2.60 - 2.70
Vanadanite	6.61 - 7.32
Zebra Stone	2.51 - 2.60

SPECIFIC GRAVITY INDEX IN ASCENDING ORDER OF MAGNITUDE

1.03 – 1.08	Copal		2.61 – 2.66	Carnelian
1.30 – 1.34	Jet		2.64 – 2.69	Aventurine
1.98 – 2.25	Opal		2.64 – 2.66	Amethyst
2.00 – 2.40	Chrysacolla		2.64 – 2.69	Mookaite
2.07 – 2.15	Sulfur		2.64 – 2.71	Tiger's Eye
2.11 – 2.16	Halite		2.65 – 2.66	Citrine
2.12 – 2.22	Stilbite		2.68 – 2.72	Labradorite
2.16 – 2.40	Scolecite		2.69 – 2.71	Iceland Spar
2.27 – 3.33	Sodalite		2.69 – 2.71	Creedite
2.28 – 2.51	Tektite		2.69 – 2.72	Sunstone
2.30 – 2.33	Selenite		2.70 – 2.71	Calcite
2.30 – 2.60	Apache Tears		2.70 – 2.76	Shiva Lingam
2.31 – 2.33	Gypsum Desert Rose		2.71 – 2.73	Herkimer Diamond
2.32 – 2.41	Apophyllite		2.71 – 2.73	Lemurian
2.40 – 2.60	Obsidian		2.71 – 2.73	Quartz
2.40 – 2.91	Lazurite		2.75 – 2.78	Emerald
2.40 – 2.91	Lapis Lazuli		2.75 – 3.00	Star Mica
2.41 – 2.62	Rainforest Rhyolite		2.80 – 2.90	Anydrite
2.32 – 2.51	Petalite		2.80 – 2.90	Angelite
2.50 – 2.65	Sepentinite		2.80 – 2.90	Lepidolite
2.50 – 2.71	Fulgarite		2.82 – 2.94	Prehnite
2.50 – 2.80	Goldstone		2.84 – 2.91	Aquamarine
2.50 – 2.91	Jasper		2.85 – 2.88	Fuchsite
2.51 – 2.60	Zebra Stone		2.85 – 2.88	Fuchsite
2.53 – 2.59	Howlite		2.90 – 3.02	Nephrite Jade
2.54 – 2.56	Charoite		2.92 – 3.03	Nuummite
2.54 – 2.59	Moonstone		2.93 – 2.94	Aragonite
2.55 – 2.63	Feldspar (orthoclase)		3.03 – 3.33	Tourmaline
2.56 – 2.59	Amazonite		3.01 – 3.33	Hypersthene
2.58 – 2.59	Chalcedony		3.10 – 3.38	Tanzanite
2.58 – 2.64	Onyx		3.17 – 3.19	Andalusite
2.58 – 2.64	Agate		3.17 – 3.19	Chiastolite
2.58 – 2.66	Pietersite		3.17 – 3.23	Apatite
2.58 – 2.91	Petrified Wood		3.18 – 3.21	Fluorite
2.59 – 2.90	Turquoise		3.20 – 3.21	Bronzite
2.60 – 2.62	Heliotrope		3.20 – 3.40	Astrophyllite
2.60 – 2.62	Bloodstone		3.20 – 4.30	Peridot
2.60 – 2.70	Chrysoprase		3.30 – 3.32	Diopside
2.60 – 2.71	Tiger Iron		3.30 – 3.40	Jadeite

3.30 – 3.50	Epidote
3.31 – 3.34	Axinite
3.35 – 3.37	Dumortierite
3.49 – 3.57	Topaz
3.40 – 3.60	Rhodonite
3.51 – 3.70	Rhodocrosite
3.53 – 3.68	Kyanite
3.70 – 4.01	Malachite
3.70 – 4.10	Green Garnet
3.80 – 4.25	Red Garnet
3.82 – 3.84	Azurite
3.86 – 4.15	Shattuckite
3.96 – 3.98	Celestite
3.96 – 3.98	Celestine
3.99 – 4.00	Corundum
3.99 – 4.00	Sapphire
3.99 – 4.01	Ruby
4.20 – 4.22	Barite
4.23 – 4.25	Rutile
4.90 – 5.20	Lodestone
4.90 – 5.20	Magnetite
4.95 – 5.10	Pyrite
4.95 – 5.10	Fool's Gold
5.08 – 5.11	Bornite
5.08 – 5.11	Chalcopyrite
5.30 – 5.33	Hematite
6.61 – 7.32	Vanadanite
7.22 – 7.61	Galena
8.96 – 8.97	Copper
9.68 – 9.74	Bismuth
10.14 – 11.00	Silver
19.30 – 19.38	Gold

MINERAL FAMILY INFORMATION

It is culturally common to mix the terms mineral, rock, and crystal. However, for mineralogy purposes the definitions are roughly as follows: "Crystal" describes any naturally occurring, inorganic, solid material, with a regular and uniform crystalline structure. Hence, all solid rocks and minerals are crystals regardless of appearance. "Minerals", are composed of singular chemical composition. "Rocks", are composed of any combination of minerals. In geology, crystal structure is a specific reference to molecular structure and pattern. In gemology and other realms, the term crystal has a loose definition and is usually based on the inherent beauty and visual appeal of faceted stones, both naturally formed or cut by man.

For the purposes of classification or for sorting a collection, rocks and minerals are commonly divided into families, based on chemical composition. Professor James Dana of Yale University developed in 1848, what has become the modern standard of classification. The Dana system divides minerals into eight basic classes.

THE CLASSES OF MINERALS ARE: **Native elements, Silicates, Oxides, Sulfides, Sulfates, Halides, Carbonates and Phosphates**. A category of **"Mineraloids"** is sometimes added to encompass organic "rocks" or compounds that cannot be classified otherwise.

The following pages list the specimens in this guide by use of the Dana system.

MINERAL FAMILY INDEX OF SPECIMENS

NATIVE ELEMENTS chemically, are molecularly pure, whereas most other minerals are made from a combination of elements. Native elements are divided into three groups; metals *(e.g. gold, silver, copper)*, semi-metals *(e.g. arsenic, antimony, bismuth)* and non-metals *(e.g. sulfur, carbon)*. Deposits of native elements are relatively rare. Although only about 20 native elements are ever readily found, they are enormously useful to us for many purposes and some are considered to be among the most valuable minerals in the world. Native elements are typically soft materials with high density.

SPECIMENS: Bismuth *(Bi-83)*, Copper *(Cu-29)*, Gold *(Au-79)*, Silver *(Ag-47)*, Sulfur *(S-16)*

SILICATES are made from metals combined with a silicate molecule *(SiO2)*. Silicates make up approximately 30% of all mineral types and almost 90% of the volume of the Earth's crust. The silicates quartz and feldspar are the most common minerals found in the world. There are six basic subfamilies of silicates, divided not by their chemistry, but by their molecular crystalline structure. They are:

NESOSILICATES, SOROSILICATES, INOSILICATES, CYCLOSILICATES, PHYLOSILICATES AND TECTOSILICATES

Because of the predominance of silicates in nature and in any collection, it is necessary and useful to divide the silicate family listing into separate indexes. A brief description of each group's crystalline structure is given at the beginning of each silicate subfamily index.

NESOSILICATES *(single tetrahedron molecular structure)*:

SPECIMENS: Andalusite *(chiastolite)*, Dumortiorite *(boro-Nesosilicates group)*, Garnet, Kyanite, Peridot, Topaz

SOROSILICATES *(double tetrahedron molecular structure)*:

SPECIMENS: Axinite *(boro-sorosilicates group)*, Epidote, Tanzanite

INOSILICATES *(single and double chain molecular structure)*:

SPECIMENS: Astrophyllite *(astrophyllite group)*, Bronzite *(orthopyroxine group)*, Diopside *(pyroxine group)*, Howlite *(boro-silicates)*, Hypersthene *(pyroxine group)*, Jadeite Jade *(pyroxine group)*, Jade Nephrite *(amphibole group)*, Nuummite *(amphibole group)*, Shattuckite *(inosilicate hydroxide)*

CYCLOSILICATES *(molecular ring structure)*:

SPECIMENS: Aquamarine *(beryl group)*, Emerald *(beryl group)*, Tourmaline *(tourmaline group)*

PHYLOSILICATES *(molecular sheet structure)*:

SPECIMENS: Apophyllite *(apophyllite group)*, Fuchsite *(mica group)*, Lepidolite *(mica group)*, Prehnite, Muscovite *(mica group)*, Star Mica *(mica group)* , Serpentinite *(serpentine group)*, Talc *(clay group)*

TECTOSILICATES *(molecular framework structure)*:

SPECIMENS: Amazonite *(orthoclase feldspar group)*, Amethyst *(quartz group)*, Citrine *(quartz group)*, Feldspar *(orthoclase feldspar group)*, Herkimer Diamond *(quartz group)*, Labradorite *(plagioclase feldspar group)*, Moonstone *(oligoclase feldspar group)*, Petalite *(feldspathoid group)*, Quartz *(quartz group)*, Scolesite *(zeolite group)*, Sodalite *(feldspathloid group)*, Stilbite *(zeolite group)*, Sunstone *(plagioclase feldspar class)*, Unakite *(quartz and orthoclase feldspar)*

CRYPTOCRYSTALLINE TECTOSILICATES *(microcrystalline)*:

SPECIMENS: Agates, Aventurine, Bloodstone *(heliotrope)*, Carnelian, Chalcedony, Charoite *(quartz chained with potassium and calcium)*, Chrysocolla *(copper silicate)*, Chrysoprase, Jasper, Lemurian Crystal, Mookaite, Onyx, Opal *(hydrated quartz)*, Petrified Wood, Pietersite, Shiva Lingam, Tektites *(fused silicate glass)*, Tiger's Eye *(silicate and pseudomorphed asbestos)*, Unakite, Zebra Stone.

MINERAL FAMILY INDEX OF SPECIMENS

OXIDES form from the combination of a metal with oxygen. They appear in a wide range of varieties from dull, common minerals, to rare gems such as rubies and sapphires. Many varieties, especially the softer ones, form close to the surface of the Earth's crust under low pressure conditions, as secondary minerals from the breakdown of sulfides and silicates. Oxides tend to have a strong chemical bond and exhibit great hardness and strength.

SPECIMENS: Corundum, Hematite, Magnetite *(lodestone)*, Rainforest Rhyolite Ruby, Rutile, Sapphire, Tiger Iron *(hematite, oxide and jasper)*

SULFIDES are made of compounds of sulfur with a metal. They are usually found as unaltered primary minerals. Upon contact with air, most transform rather rapidly into an oxide mineral. They tend to be heavy and brittle. Sulfides are the main source of most useful metal ores. They are usually high density, and unlike pure metals are brittle.

SPECIMENS: Bornite *(peacock ore)*, Galena, Pyrite

SULFATES are made of a metal combined with a sulfur and oxygen molecule *(SO4)*. This is a large group of minerals that tend to be soft, and many are semi-translucent. They commonly occur as evaporates from sea water or in hydrothermal veins, and can be a secondary mineral when a sulfide mineral oxidizes.

SPECIMENS: Anhydrite *(angelite)*, Barite, Celestite *(celestine)*, Selenite

HALIDES form from a metal in combination with a halogen element such as chlorine, bromine, fluorine or iodine. Halides are very soft and water soluble. They are extremely abundant and many find usage in daily life, such as halite *(NaCl)* which is common table salt. Halides are usually found in pegmatite intrusions, in igneous or metamorphic rock beds.

SPECIMENS: Creedite, Fluorite, Halite

CARBONATES are a group of minerals made of a metal and the carbonate ion *(CO3-2)*. Calcite, also known as calcium carbonate, is the most common and basic of the carbonate group and is an abundant contributor to the formation of limestone. Carbonates are among the most widely distributed groups of minerals in the Earth's crust. Carbonates are common as evaporates in dry lake beds or in association with hydrothermal vents.

SPECIMENS: Aragonite, Azurite, Calcite, Iceland Spar *(calcite)*, Malachite, Rhodocrosite

PHOSPHATES are inorganic salts composed of a metal and phosphoric acid molecule *(H3PO4)* and are fairly uncommon. Primary phosphates are always liquid-formed and found in igneous rock formations. Secondary phosphates are altered forms of minerals, commonly found in low pressure, low temperature zones close to the Earth's surface. Phosphates are generally soft, and often brightly colored. Commercially, phosphates are used mostly for fertilizers.

SPECIMENS: Apatite, Lazurite *(lapis lazuli)*, Turquoise, Vanadanite

MINERALOIDS is the term used for substances that do not fit into one of the eight classes.

ORGANICS: Amber *(fossilized tree sap)*, Copal *(not-yet fossilized tree sap)*, Jet *(fossilized plant matter)*

GLASSES: Apache Tears *(volcanic)*, Fulgarite *(lightning-fused)*, Goldstone *(manufactured)*, Obsidian *(volcanic)*, Opalite *(manufactured)*

CONCRETIONS: Boji Stone *(pyrite and mixed minerals)*, Moqui Marble *(hematite surrounding sandstone)*, Septarian *(aragonite, bentonite, limestone)*

CLASSIFICATION BY CRYSTAL SYSTEM

The common method of classifying crystals is by their atomic lattice or molecular structure. A lattice is a three dimensional framework of atoms arranged in a symmetrical pattern. The lattice shape determines which crystal system a stone belongs to, and is also the basis for its physical properties and appearance. There are seven crystal systems or groups, however hexagonal crystals are divided into two subgroups; hexagonal and trigonal hexagonal. In addition to the systems defined by crystal structure, there is an additional group for amorphous minerals, which have no crystalline structure.

The index of specimens on the following pages are categorized according to the following crystal systems:

ISOMETRIC, HEXAGONAL, TRIGONAL HEXAGONAL, TETRAGONAL, ORTHORHOMBIC, MONOCLINIC, TRICLINIC, AMORPHOUS

SPECIMENS WITH ISOMETRIC CRYSTAL STRUCTURE
The isometric system is also known as the cubic system. Isometric crystals have three axes of equal length and intersect at right angles. Overall structure is based on a square inner structure. Isometric crystal shapes include: cube, octahedron, rhombic dodecahedron, icositetrahedron and hexicisohedron.

ISOMETRIC CRYSTALS: Copper, Fluorite, Fool's Gold, Galena, Garnet, Gold

SPECIMENS WITH HEXAGONAL CRYSTAL STRUCTURE
Hexagonal crystals have three out of their four axes in one plane, are of the same length, and intersect each other at angles of 60 degrees. The fourth axis is of a different length and intersects the others at right angles. Overall structure is based on a hexagonal inner structure. Hexagonal crystal shapes include: four-sided prism and pyramid, twelve-sided pyramid and double pyramid.

HEXAGONAL CRYSTALS: Apatite, Aquamarine, Corundum, Emerald, Ruby, Sapphire, Serpentinite, Vanadanite

SPECIMENS WITH TRIGONAL HEXAGON CRYSTAL STRUCTURE

Axes and angles in this system are similar to the hexagonal system, and the two systems are often combined as hexagonal. In the cross-section of a hexagonal crystal, there will be six sides. In the cross-section of a trigonal crystal there will be three sides. Overall structure is based on a triangular inner structure. Crystal shapes include: three-sided prism or pyramid, rhobohedron and scalenohedron.

TRIGONAL CRYSTALS: Agate, Amethyst, Chalcedony, Citrine, Jasper, Lemurian Crystal, Mookaite, Onyx, Petrified Wood, Pietersite, Quartz, Rainforest Rhyolite, Rhodocrosite, Tiger's Eye, Zebra Stone

SPECIMENS WITH TETRAGONAL CRYSTAL STRUCTURE

Tetragonal crystals have two axes of equal length and are in the same plane, the main axis is either longer or shorter and all three intersect at right angles. Overall structure is based on a rectangular inner structure. Crystal shapes include: four-sided prism, pyramid, trapezohedron, eight-sided and double pyramid, icositetrahedron and hexicisohedron.

TETRAGONAL CRYSTALS: Apophyllite, Garnet, Rutile

SPECIMENS WITH ORTHORHOMBIC CRYSTAL STRUCTURE

Orthorhombic crystals have three axes, all of different lengths and are at right angles to each other. Overall structure is based on a rhombic *(diamond-shaped)* inner structure. Crystal shapes include: pinaclid, rhombic, prism, pyramid and double pyramid.

ORTHORHOMBIC CRYSTALS: Andalusite, Angelite, Anhydrite, Aragonite, Barite, Bornite, Bronzite, Celestine, Celestite, Chiastolite, Chrysocolla, Hypersthene, Nuummite, Peridot, Prehnite, Shattuckite, Sulfur, Tanzanite, Topaz

CLASSIFICATION BY CRYSTAL SYSTEM CON'T

SPECIMENS WITH MONOCLINIC CRYSTAL STRUCTURE
Monoclinic crystals have three axes, each of different lengths. Two are at right angles to each other and the third is inclined. Overall structure is based on a parallelogram inner structure. Crystal shapes include: basal pinacoid and prism with inclined end faces.

MONOCLINIC CRYSTALS: Azurite, Bloodstone, Chalcedony *(trigonal monoclinic)*, Charoite, Chrysoprase, Creedite, Diopside, Epidote, Fuchsite

SPECIMENS WITH ORTHORHOMBIC CRYSTAL STRUCTURE
Orthorhombic Crystals have three axes, all of different lengths and all at right angles to each other. Overall structure is based on a rhombic *(diamond-shaped)* inner structure. Crystal shapes include: pinacoid, rhombic, prism, pyramid and double pyramid.

ORTHORHOMBIC CRYSTALS: Andalusite, Angelite, Anhydrite, Aragonite, Barite, Bornite, Bronzite, Celestine, Celestite, Chiastolite, Chrysocolla, Hypersthene, Nuummite, Peridot, Prehnite, Shattuckite, Sulfur, Tanzanite, Topaz

SPECIMENS WITH MONOCLINIC CRYSTAL STRUCTURE
Monoclinic Crystals have three axes, each of different lengths. Two are at right angles to each other and the third is inclined. Overall structure is based on a parallelogram inner structure. Crystal shapes include: basal pinacoid and prism with inclined end faces.

MONOCLINIC CRYSTALS: Azurite, Bloodstone, Chalcedony *(trigonal monoclinic)*, Charoite, Chrysoprase, Creedite, Diopside, Epidote, Fuchsite, Gypsum Desert Rose, Howlite, Jadeite, Jade Nephrite, Lepidolite, Malachite, Star Mica, Muscovite, Moonstone, Petalite, Scolesite, Selenite, Stilbite

SPECIMENS WITH TRICLINIC CRYSTAL STRUCTURE

Triclinic crystals have all three axes of different lengths and are inclined towards each other. Crystal forms are usually paired faces.

TRICLINIC CRYSTALS: Amazonite, Astrophyllite, Axinite, Carnelian, Kyanite, Labradorite, Rhodonite, Sunstone, Turquoise

AMORPHOUS ROCKS & MIXED MINERAL SPECIMENS

Amorphous minerals have no organized crystalline structure. Mixed minerals contain more than one crystalline structure and so they are not categorized in this guide by crystal structure.

AMORPHOUS MINERALS: Amber, Apache Tears, Copal, Fulgarite, Goldstone, Jet, Obsidian, Opal, Shungite, Tektite

MIXED MINERALS: Boji Stone, Moqui Marble, Ruby in Fuchsite, Septarian, Tiger Iron, Unakite

INTRODUCTION TO CRYSTAL HEALING

The following index is meant to serve as a basic guide for those interested in learning about spiritual uses for their stones, and also for those whom are currently knowledgeable about crystal healing. Crystal therapy and other healing methods of this sort are certainly not scientific in basis, but some aspects have been found by some, to have possible therapeutic effects. For healing any ailment we recommend seeking professional medical advice.

The information in this book is not intended as a replacement for medical treatment.

We do believe that no matter the chosen method of healing, the power of the mind, intention, and positive thought, are certainly beneficial to any healing process whether it be physical, emotional, or spiritual in nature. Our purpose is not to promote, to prove, or to disprove, any aspect of these practices. This guide is simply meant to be an easy reference to as many uses for healing crystals as possible. We have tried diligently to rely on only trusted sources when compiling this index. We realize that there are many different opinions on specific stones for specific purposes. Some crystal healing practices are ancient, sometimes hundreds or even thousands of years old. Thus, there are apt to be many different interpretations and variations of these practices as they apply to modern times. While some opinions might be considered contradictory, we feel it is important to remember that here is no "wrong" way to practice spiritual healing. When necessary, we have used our best judgement in this book when decisions about conflicting information had to be made.

As with all healing practices, there is certainly room for personal interpretation with crystal healing, and no particular usage or method mentioned in this book is "set in stone". You are encouraged to practice and experiment with your crystals and stones. Any stone or crystal can be of benefit to you, as long as you are comfortable with it, are drawn to it, or just plain like it!

Please, take some time to cross-reference the crystals from the healing index with the geologic and mineralogical information in this book. We feel it is important to have as complete an understanding of a crystal as possible when embarking on any spiritual endeavor.

METAPHYSICAL & HEALING GUIDE—
INDEXED BY AILMENT OR USAGE

A

ABDOMINAL PAIN
Blue Lace Agate, Garnet, Acid indigestion

ACID INDIGESTION
Apatite, Carnelian, Diopside, Clear Quartz, Rhodocrosite, Peridot, Bismuth

ACNE
Amethyst, Citrine

ADD/ADHD
Amethyst, Astrophyllite, Petalite, Smoky Quartz

ADDICTION
White Agate, Chevron Amethyst, Fluorite, Labradorite, Lepidolite, Clear Quartz, Ruby Fuchsite

ADRENALINE
Green Quartz

AGING
Chiastolite, Citrine

AIR PURIFICATION
Halite, Selenite

AIR SICKNESS
Malachite, Jet, Sapphire, Pink Tourmaline

AKASHIC RECORDS (ACCESS)
Apophyllite, K2 Stone, Lepidolite, Quartz Phantom

ALLERGIES (AIRBORNE)
Apatite, Aquamarine, Aventurine, Brecciated Jasper, Sodalite, Unakite

ALLERGIES (FOOD)
Apatite, Calcite, Citrine, Fuchsite, Lepidolite, Prehnite, Picture Jasper, Yellow Jasper, Sulfur

ALZHEIMER'S
Lepidolite, Chalcedony, Rose Quartz, Galena

AMPLIFY
(other stones & crystals), Herkimer Diamond, Jet, Clear Quartz

ANEMIA
Bloodstone, Copper, Tangerine Quartz, Ruby

ANGER (INJURY OR SWELLING)
Blue Lace Agate, Aventurine, Star Mica, Yellow Apatite

ANIMALS (PROTECTION)
Green Apatite, Serpentinite

ANIMALS (TEMPERAMENT & HEALING)
Brecciated Jasper, Leopardskin Jasper, Chiastolite

ANKLES (INJURY OR SWELLING)
Moss Agate, Dalmatian Jasper

ANOREXIA
Topaz

ANTIBACTERIAL
Copper, Silver

ANTI DEPRESSION
Botswana & Pink Botswana Agate, Yellow Apatite, Orangeskin Jasper

ANTI INFLAMMATORY
Angelite, Blue Lace Agate, Turquoise,

ANXIETY
Hematite, Rose Quartz, Rainforest Rhyolite

APPETITE
Star Mica

ARRHYTHMIA
Blue Agate, Bloodstone,
Green Obsidian, Green Quartz,
Rhodochrosite

ART CREATIVITY
Aventurine, Carnelian, Chrysacolla,
Brecciated Jasper, Moonstone,
Tiger Iron

ARTERIES
Anhydrite (Angelite)

ARTHRITIS
Iceland Spar, Amazonite, Carnelian,
Moss Agate, Chrysacolla, Rhodonite

ASTHMA
Amber, Azurite, Lazurite, Malachite,
Carnelian

ASTRAL TRAVEL
Yellow Apatite, Apophyllite, Dogtooth
Calcite, Fulgarite, Kambaba Jasper

ATTACK (PSYCHIC)
Hypersthene, Apophyllite,

Aura (Cleanse)
Pink Botswana Agate, Jet, Tektite,
Zebra Stone

AUTO-IMMUNITY
Rhodonite

**AWAKENING
(SPIRITUAL, KUNDALINI)**
Use a full set of chakra stones of your
own liking.

B

BACK PAIN
Amazonite, Azurite, Yellow Calcite,
Carnelian, Petrified Wood

BAD HABITS
Astrophyllite, Labradorite, Lepidolite

BAD MOOD
Blue Lace Agate, Sodalite, Ruby,
Smoky Quartz

BALANCE (EMOTIONAL)
Copper, Peruvian Agate, Boji Stone,
Desert Jasper, Lepidolite, Star Mica

BED WETTING
Amazonite, Carnelian, Rhodochrosite,
Rose Quartz, Sodalite, Tiger's Eye

BIRTHING
Unakite, Onyx, Opal, Picture Jasper

BLADDER ISSUES
Apophyllite, Bloodstone, Citrine

BLEEDING
Rhodonite, Hematite, Bloodstone,
Rose Quartz, Lodestone

BLOOD PRESSURE
Aventurine, Blue Calcite, Lazurite,
Malachite, Rose Quartz

BODILY FLUIDS
Blue Obsidian, Tree Agate

BONE HEALTH
Blue Apatite, Iceland Spar, Goldstone,
Zebra Jasper, Onyx

BONE MARROW
Chrysacolla, Malachite

BROKEN BONES
Axinite, Gold, Tiger's Eye,
Hematite, Malachite

BRUISES
Amber, Amethyst, Sodalite, Lazurite

BULIMIA
Topaz, Garnet, Feldspar, Rose Quartz

BURNS
Amethyst

BURSITIS
Iceland Spar, Amazonite, Carnelian,
Moss Agate, Chrysacolla, Goldstone,
Rhodonite

C

**CALCIUM
(DEFICIENCY, ABSORPTION)**
Howlite, Orange Calcite, Iceland Spar,
Shattuckite

CALMING
Blue Lace Agate, Aquamarine, Blue
Calcite, Howlite

CANCER
Malachite, Prehnite, Lapis,
Sodalite (do not use Sodalite
during chemotherapy)

CAR SICKNESS
Malachite, Jet, Sapphire, Pink
Tourmaline

CARTILAGE
Dalmatian Jasper, Sunstone

CATARACTS
Apophyllite, Blue Obsidian

CELLULITE
Astrophyllite, Yellow Apatite, Citrine

CELLS (DAMAGE/REPAIR)
Apophyllite, Garnet, Malachite,
Snowflake Obsidian, Tanzanite

CHAKRA (CLEANSE, ALIGN)
Charoite, Copper, Garnet, Smoky
Quartz, Shiva Lingam

CHANGE
Axinite, Chevron Amethyst, Bismuth

CHEMOTHERAPY
Prehnite, Smoky Quartz, Selenite,
Tanzanite, Vanadanite

CHILDBIRTH
Unakite, Onyx, Opal, Picture Jasper

CHILDREN (PROTECTION)
Yellow Obsidian

CHILLS
Opal, Black Obsidian

CHROMOSOME (DAMAGE)
Apophyllite, Garnet, Malachite,
Snowflake Obsidian, Chiastolite

CHRONIC CONDITIONS
Apophyllite, Diamond, Clear Quartz

CIRCULATION
Moss Agate, Pink Botswana Agate,
Bloodstone, Red Jasper, Lazurite

CLAIRVOYANCE
Apophyllite, Emerald, Purple Fluorite,
Kyanite, Labradorite, Nuummite

CLARITY (MENTAL)
White Agate, Aquamarine, Celestite,
Kambaba Jasper, Boji Stone

CLEAN AIR
Halite, Selenite

CLEAN WATER
Lodestone, Shungite

CLEANSING
Amethyst, Citrine, Copal, Emerald, Jet,
Peridot, Sugilite

CLEANSING (OTHER CRYSTALS)
Apophyllite, Jet, Selenite, Scolecite,
Stilbite

CLUMSINESS
Muscovite (Mica)

CODE BREAKING
Nuummite

CODEPENDENCY
Fuchsite

COLD (HEAD)
Opal, Amber, Carnelian, Rainbow
Fluorite, Jet, Sulfur

**COLLEGE
(STARTING OR RETURNING TO)**
Diopside, Malachite

COLIC
Boji Stones, Amber

COLON
Apophyllite, Orange Calcite

COMMUNICATION
Chrysacolla, Emerald, Blue Fluorite,
Fulgarite, Geode, Moqui Marble,
Septarian

COMPASSION
Aventurine, Staurolite

CONCENTRATION
Malachite, White Agate, Amazonite,
Boji Stone, Celestite, Kambaba Jasper,
Clear Quartz, Sapphire

CONFIDENCE
Blue Apatite, Moss Agate, Bronzite,
Gold, Goldstone, Ruby

CONFUSION
Green Apatite

CONSTIPATION
Citrine, Orangeskin Jasper

COOLING
Blue Lace Agate, Dumortierite

COOPERATION
Feldspar, Green Quartz, Ruby Fuchsite

COORDINATION
Copper, Green Apatite, Black
Tourmaline

COUGH
Pyrite, Amber, Aquamarine, Topaz

COURAGE
Aquamarine, Carnelian

CRABBINESS
Blue Lace Agate, Sodalite, Ruby,
Smoky Quartz

CRAMPS (MOON-CYCLE)
Garnet, Black Obsidian, Chrysocolla,
Moonstone, Jade

CRAMPS (MUSCLE)
Turquoise, Bloodstone, Blue Apatite,
Aragonite, Creedite, Orangeskin
Jasper

CREATIVE BLOCKS
Orange Calcite, Chiastolite

CREATIVITY
Aventurine, Carnelian, Chrysacolla,
Brecciated Jasper, Moonstone, Tiger
Iron

CROHN'S DISEASE
Green Fluorite, Smoky Quartz, Rose
Quartz, Yellow Jasper, Carnelian,
Turritella Agate, Pyrite

CRYSTALS (AMPLIFY)
Herkimer Diamond, Jet, Clear Quartz

CURSES (REMOVE)
Hypersthene, Labradorite, Sulfur

CUTS
Tangerine Quartz, Ruby, Blue Lace,
Halite, Jade, Rhodonite

D

DNA (DAMAGE/REPAIR)
Apophyllite, Garnet, Malachite,
Snowflake Obsidian, Chiastolite

DAYDREAMING
Snowflake Obsidian

DECISION MAKING/
DECISIVENESS
Red Obsidian, Moonstone, Onyx

DEHYDRATION
Brecciated Jasper, Mookaite, Blue
Obsidian, Moss Agate

DEMENTIA
Chalcedony, Lepidolite, Kyanite,
Lepidolite, Blue Agate

DEPRESSION
Botswana & Pink Botswana Agate,
Yellow Apatite, Orangeskin Jasper

DETOXIFICATION
Yellow Apatite, Azurite, Celestite, Red
Jasper, Dumortierite

DIABETES
Shattuckite, Jade, Citrine

DIARRHEA
Black Tourmaline, Smoky Quartz

DIGESTION
Yellow Apatite, Bismuth

DIZZINESS
Pietersite

DREAMS (PLEASANT)
Amethyst, Charoite, Herkimer
Diamond, Hematite

DREAMS
(RECALL AND VIVID DREAMING)
Yellow Apatite, Red Calcite, Celestite

DRIVING (LEARNING, TESTING)
Diopside, Kyanite

DYSLEXIA
Black Tourmaline, Diopside

E

EARACHE
Purple Agate, Amethyst, Celestite,
Sapphire

EARTH (ECOLOGY)
Turritella Agate, Desert Jasper, Picture
Jasper, Green Obsidian, Zebra Stone

EATING DISORDERS
Topaz, Unakite

ECOLOGY
Turritella Agate, Desert Jasper, Picture
Jasper, Green Obsidian, Zebra Stone

ECZEMA
Sapphire, Picture Jasper

ELECTROMAGNETIC (EM)
POLLUTION
Red Jasper, Sodalite, Tourmalinated
Quartz

EMOTIONAL BALANCE
Peruvian Agate, Boji Stone, Desert
Jasper, Lepidolite, Star Mica

EMOTIONAL PAIN
Crazy Lace Agate, Howlite

EMPHYSEMA
Rhodonite

ENDOCRINE SYSTEM/
GLANDS-GREEN
Quartz, Dumortierite

ENERGY
Boji Stone, Bumblebee , Citrine,
Copper, Fancy Jasper

ENERGY (NEGATIVE)
Chevron Amethyst, Apache Tears,
Bronzite, Nuummite, Black Obsidian

ENLIGHTENMENT
Botswana Agate

EPILEPSY
Malachite

EXPRESSION (SELF)
Chrysacolla, Emerald, Blue Fluorite,
Fulgarite, Moqui Marble

EYES
Peruvian Agate, Aventurine, Celestite,
Chalcedony, Sapphire, Picasso Jasper

F

FARMING
Any Agate, Jasper or Feldspar

FASTING
Star Mica

FATIGUE
Orange Calcite, Bronzite, Marcasite

FEAR (GENERAL)
Aquamarine, Amazonite, Iceland Spar

FEAR OF WATER
Blue Obsidian, Ocean Jasper

FEET
Turritella Agate, Onyx

FERTILITY
Shiva Lingam

FEVER
Blue Lace gate, Jet, Bornite, Clear
Quartz, Tangerine Quartz, Sapphire,
Sodalite

FIBROMYALGIA
Amber, Jade, Moonstone,
Tourmalinated Quartz

FINDING LOST ITEMS
Snakeskin Agate, Turritella Agate,
Nuummite

FINGERS
Moss Agate

FINGER NAILS
Pink Tourmaline, Onyx, any Calcite

FLU
Peacock Ore, Tourmaline, Sodalite

FLUIDS (BODILY)
Blue Obsidian, Tree Agate

FOCUS
Crazy Lace Agate, Malachite, White
Agate, Boji Stone, Celestite, Kambaba
Jasper, Clear Quartz, Sapphire

FOOD ALLERGIES
Calcite, Citrine, Fuchsite, Lepidolite,
Prehnite, Picture Jasper, Sulfur

FOOT CRAMPS
Hematite, Lodestone, Magnetite,
Tiger Iron

FRACTURES (BONES)
Gold Tiger's Eye, Hematite, Malachite

FRIENDSHIP
Aventurine, Turquoise, Picasso Jasper,
Yellow Jasper

FRISBEE (AND DISC GOLF)
Discusite, Whammo-ite

FRUSTRATION
Blue Obsidian

G

GALL BLADDER
Chalcedony

GARDEN PLANTS
Any Agate, Jasper or Feldspar and
Mookaite

GASTROENTERITIS & IBD
Green Fluorite, Carnelian, Turritella
Agate

GENERAL HEALING
Boji Stone, Fancy Jasper, Mookaite,
Smoky Quartz

GENEROSITY
Topaz

GLANDS
Aquamarine

GOALS (ACHIEVING)
Carborundum , Feldspar

GRIEF
Jet, Rhodonite, Moonstone, Lepidolite,
Onyx

GROUNDING
Peruvian Agate, Smoky Quartz

GROWTH (SPIRITUAL)
Unakite

GUILT
Fancy Jasper

GUMS (& TEETH)
Any Fluorite, Zebra Jasper, Onyx,
Sunstone, Zebra Stone

H

HAIR
Petrified Wood, Pink Tourmaline,
Tanzanite

HANDS
Turritella Agate

HANGOVER
Aquamarine, Chrysocolla, Turquoise,
Sodalite, Hematite, Petrified Wood

HAPPINESS
Dalmatian Jasper

HEADACHE
Purple Agate, Amethyst, Aventurine,
Snowflake Obsidian

HEAD COLD
Opal, Amber, Carnelian, Rainbow
Fluorite, Jet, Sulfur

HEALING (GENERAL)
Boji Stone, Fancy Jasper, Mookaite,
Smoky Quartz

HEALING (SELF)
Jade, Copal, Leopardskin Jasper

HEART
Blue Agate, Blue Calcite, Creedite,
Serpentine, Zebra Stone

HEARTBURN
Apatite, Carnelian, Diopside, Clear
Quartz, Rhodocrosite, Peridot, Bismuth

HEART PALPITATIONS
Blue Agate, Bloodstone, Green
Obsidian, Green Quartz

HERPES
Jadeite, Lazurite

HEXES
Hypersthene, Labradorite, Sulfur

HOARSENESS
Amber, Sodalite, Lazurite

HOUSEHOLD PROTECTION
Creedite, Carnelian

HOT FLASHES
Moonstone, Chrysocolla

HOUSEPLANTS
Any Agate or Jasper or Feldspar

HUNGER
Star Mica

HYPERACTIVITY
Dogtooth Calcite, Dumortierite

HYPOTHYROIDISM
Blue Tourmaline, Aquamarine,
Angelite

I

IRRITABLE BOWEL (IBD)
Green Fluorite, Carnelian, Turritella
Agate

IMMUNE SYSTEM
Epidote, Apache Tears, Moss Agate,
Creedite, Desert Jasper, Mookaite

IMPOTENCE/INFERTILITY
Shiva Lingam

INDIGESTION
Apatite, Carnelian, Diopside, Clear
Quartz, Rhodocrosite, Peridot,
Bismuth

INDIGO CHILD
Pink Tourmaline

INFECTION
Blue Lace Agate, Copper, Sulfur, Jet,
Malachite

INFLAMMATION
Angelite, Blue Lace Agate, Turquoise

INNER PEACE
Apophyllite, Carborundum , Scolecite

INSECT BITES
Turritella Agate

IRRITABILITY
Plagioclase Feldspar

INSOMNIA
Amethyst, Red Calcite, Celestite, Blue
Fluorite, Hypersthene, Ruby Fuchsite

INTELLECT
Yellow Calcite, Emerald

INTERNAL CLOCK (& JET LAG)
Bronzite, Hematite, Black Tourmaline

INTERNAL ORGANS
Crazy Lace Agate

INTUITION
Purple Agate, Amethyst, Apophyllite,
Emerald, Gypsum Desert Rose,
Labradorite

IRON DEFICIENCY
Bronzite

ITCH
Turritella Agate

J

JET LAG
Bronzite, Hematite, Black Tourmaline

JOINTS
Iceland Spar, Amazonite, Carnelian, Moss Agate, Chrysacolla, Goldstone, Rhodonite

JOY
Dalmatian Jasper

K

KIDNEYS
Bloodstone, Orange Calcite, Carnelian, Desert Jasper, Rainforest Rhyolite, Prehnite

KINDNESS
Blue Agate

KUNDALINI ENERGY & AWAKENING
Serpentine, Shiva Lingam or a full set of Chakra stones.

L

LANGUAGE (LEARNING)
Red Jasper, Topaz

LARYNGITIS/LARYNX
Amber, Sodalite, Lazurite

LAUGHTER
Ruby

LEARNING
Yellow Apatite, Blue Calcite, Kambaba Jasper, Malachite, Star Mica, Clear Quartz

LEG CRAMPS
Hematite, Magnetite, Howlite

LIGHTNING & THUNDER (FEAR)
Moss Agate, Hypersthene

LIVER
Azurite, Bloodstone, Chrysacolla

LOST ITEMS SNAKESKIN AGATE,
Turritella Agate, Nuummite

LOVE
Emerald, Garnet, Gold, Prehnite, Rose Quartz, Ruby

LUCK
Dalmatian Jasper, Gold, Copal, Nuummite, Sunstone, Gold Tiger's Eye

LEUKEMIA
Bloodstone, Chalcedony, Chrysacolla

LUNGS
Blue Agate, Amethyst, Apophyllite, Aventurine, Green Quartz, Rutile Quartz, Sugilite

M

MAGIC
Lodestone, Nuummite, Labradorite, Sulfur

MEDITATION
Purple Agate, Herkimer Diamond, Hypersthene, Lepidolite, Pinolite

MEDITATION (OUTDOORS)
Quartz (fire), Blue Obsidian (water), Apache Tears (earth), Sunstone (air)

MEMORY
Yellow Apatite, Blue Calcite, Kambaba Jasper, Malachite, Star Mica, Clear Quartz

MENOPAUSE
Diopside, Lepidolite

MENTAL FUNCTION
Peruvian Agate, Botswana Agate,
Kambaba Jasper, Malachite, Star Mica

METABOLISM
Aventurine, Dogtooth Calcite, Garnet,
Bornite, Blue Tiger's Eye

MIGRAINE
Purple Agate, Amethyst, Aventurine,
Snowflake Obsidian

MISTAKES (MENDING)
Picture Jasper, Green Obsidian

MOODINESS
Blue Lace Agate, Sodalite, Ruby,
Smoky Quartz

MOOD SWINGS
Blue Tiger's Eye

**MOON CYCLES
(REGULATION & CRAMPS)**
Amber, Bloodstone, Chrysocolla,
Garnet, Hematite, Moonstone, Jade,
Turquoise

MOTION SICKNESS
Malachite, Jet, Sapphire, Pink
Tourmaline

MORNING
Jet, Rhodonite, Moonstone, Lepidolite,
Onyx

MOUTH
Red Obsidian

MUCOUS (EXCESS)
Angelite, Pink Tourmaline, Aventurine,
Chrysocolla, Pietersite, Prehnite

MUMPS
Aquamarine, Topaz

MULTIPLE SCLEROSIS
Lazurite, Sodalite, Blue Sapphire,
Moonstone, Amethyst

MUSCLES
Apache Tears, Blue Apatite, Howlite

MUSCLE SPRAINS
Dalmatian Jasper

MUSCLE SPASMS
Turquoise, Bloodstone, Blue Apatite,
Aragonite, Creedite, Orangeskin
Jasper

MUSICAL CREATIVITY
Carnelian, Chrysacolla, Brecciated
Jasper, Moonstone, Tiger Iron

N

NAILS
Apatite, Pink Tourmaline, Onyx, Any
Calcite

NAUSEA
Emerald, Lazurite, Plagioclase
Feldspar

NATURE
Any Agate, Jasper or Feldspar

NEGATIVE ENERGY (REMOVE)
Chevron Amethyst, Apache Tears,
Bronzite, Nuummite, Black Obsidian

NERVES/NERVOUS SYSTEM
Amazonite, Creedite, Pinolite

NEURALGIA
Carnelian, Lepidolite, Tree Agate,
Lazurite

NIGHT VISION
Amber, Fire Agate, Tiger's Eye

NIGHTMARES
Charoite, Amethyst, Herkimer
Diamond, Hematite

NOSEBLEEDS
Rhodonite, Hematite, Bloodstone,
Rose Quartz, Lodestone

O

ORGANS (INTERNAL)
Crazy Lace Agate

ORGANIZATION
Bloodstone, Dumortiorite, Brecciated
Jasper, Fluorite, Lazurite (Lapis Lazuli)

OXYGENATION
Blue Agate, Pink Botswana Agate

P

PAIN
Amethyst, Aragonite, Crazy Lace
Agate, Howlite, Kyanite, Moqui Marble

PAIN (ABDOMINAL)
Blue Lace Agate, Garnet

PALPITATIONS
Blue Agate, Bloodstone, Green
Obsidian, Green Quartz

PANCREAS
Citrine, Zircon

PARKINSON'S DISEASE
Opal

PAST ISSUES
Tangerine Quartz, Opal, Nuummite

**PAST LIFE
(CONNECTION & REGRESSION)**
Creedite, Apophyllite, Preseli
Bluestone, Serpentine, Tibetan Quartz

PATIENCE
Blue Agate, Silver

PEACE
White Agate, Angelite, Creedite,
Lazurite, Scolecite, Super 7

PERSONAL POWER
Amazonite

PETS (PROTECTION)
Green Apatite

PETS (TEMPERAMENT & HEALING)
Brecciated Jasper, Leopardskin
Jasper, Chiastolite

PLANET (ECOLOGY)
Turritella Agate, Desert Jasper, Picture
Jasper, Green Obsidian, Zebra Stone

PLANTS
Any Agate, Jasper or Feldspar

PMS
Moonstone, Jade, Chrysocolla,
Turquoise

PNEUMONIA
Apophyllite, Bloodstone, Any Fluorite,
Galena, Rose Quartz.

POLLUTION (ELECTROMAGNETIC)
Red Jasper, Tourmilinated Quartz

PREGNANCY
Unakite, Onyx, Opal, Picture Jasper

PROBLEM SOLVING
Rutile Quartz, Topaz

PROSPERITY
Aventurine, Orange Calcite, Citrine,
Jade

PROTECTION
Children

PROTECTION (GENERAL)
Amethyst, Bloodstone, Nuummite,
Snowflake Obsidian

PROTECTION (HOUSEHOLD)
Creedite, Carnelian

PSYCHIC ATTACK (PROTECTION)
Lazurite, Black Obsidian,
Tourmalinated Quartz, Black
Tourmaline

PSYCHIC ABILITY
Apophyllite, Emerald, Purple Fluorite,
Kyanite, Labradorite, Nuummite

PSYCHIC VISION
Unakite

PUBLIC SPEAKING
Chrysacolla, Emerald, Blue Fluorite,
Fulgarite, Moqui Marble, Septarian

PURIFICATION (AIR)
Halite, Zircon

PURIFICATION (WATER)
Lodestone

R

**RADIATION
(RECOVERY & PROTECTION)**
Yellow Agate, Prehnite, Selenite,
Tanzanite, Vanadanite

RASH
Turritella Agate

RECOVERY (ILLNESS)
Epidote, Fuchsite, Jade, Brecciated
Jasper, Moss Agate

REIKI
Apophyllite, Stilbite, Clear Quartz,
Sunstone

RELAXATION
Amethyst, Larimar, Lepidolite

REPRESSION
Pink Botswana Agate

REPRODUCTIVE HEALTH
Gold Tiger's Eye

RESPIRATORY
Blue Agate, Amethyst, Apophyllite,
Aventurine, Green Quartz, Rutile
Quartz

RIB CAGE
Azurite

ROMANCE
Emerald, Garnet, Gold, Prehnite, Rose
Quartz, Ruby

RUDENESS (DEALING WITH)
Hematite, Howlite, Rhodonite

S

SADNESS
Apache Tears

SAFE TRAVEL
Agates, Moonstone

SCIATICA
Lepidolite

SCRAPES
Unakite, Onyx, Opal, Picture Jasper,
Tangerine Quartz

SECRETS (SOLVING)
Nuummite

SELF ESTEEM
Anyolite, Blue Apatite, Moss Agate, Bronzite, Gold, Goldstone, Ruby

SELF EXPRESSION
Chrysacolla, Emerald, Blue Fluorite, Fulgarite, Moqui Marble

SELF HEALING
Jade, Copal, Leopardskin Jasper

SENILITY
Lepidolite, Kyanite, Lepidolite, Blue Agate

SEROTONIN (BALANCE)
Scolecite, Halite

SHYNESS
Orange Calcite, Garnet, Topaz, Tiger's Eye, Jade, Opal, Obsidian

SINUSES
Aventurine, Opal, Scolecite

SINCERITY
Amazonite

SKIN
Pink Botswana Agate, Azurite, Desert Jasper, Pet. Wood, Rhodochrosite, Serpentine

SLEEP
Amethyst, Red Calcite, Celestite, Blue Fluorite, Hypersthene, Ruby Fuchsite

SMOKING (QUITTING)
Feldspar, Diopside, Fluorite, Labradorite, Quartz, Ruby

SNAKEBITES
Snakeskin Agate

SNORING
Pyrite

SORE THROAT
Angelite, Celestite, Copal, Blue & Rainbow Fluorite, Lazurite, Sodalite

SORROW
Apache Tears, Obsidian, Prehnite

SPASMS (MUSCLE)
Turquoise, Bloodstone, Blue Apatite, Aragonite, Creedite, Orangeskin Jasper

SPEECH/PUBLIC SPEAKING
Chrysacolla, Emerald, Blue Fluorite, Fulgarite, Larimar, Moqui Marble

SPINE
Azurite, Axinite, Yellow Calcite, Carnelian, Petrified Wood

SPIRIT GUIDES
Angelite, Leopardskin Jasper

SPIRITUAL GROWTH
Larvikite, Unakite

SPLEEN
Bloodstone, Chalcedony, Citrine

SPRAINS (MUSCLE)
Dalmatian Jasper

STAMINA
Boji Stone, Citrine, Copper, Fancy Jasper

STONES (AMPLIFY)
Herkimer Diamond, Jet

STRENGTH (PHYSICAL)
Hematite, Marcasite

STRESS
Aragonite, Bornite, Fuchsite, Desert Rose, Hypersthene, Petalite

STUDYING
Yellow Apatite, Blue Calcite, Kambaba Jasper, Malachite, Star Mica, Clear Quartz

SUCCESS
Aventurine, Silver, Thulite

SUNBURN
Amethyst

SURGERY
Diopside

SWELLING
Hematite, Tourmaline

T

TEETH (& GUMS)
Any Fluorite, Zebra Jasper, Onyx, Stromatolite, Sunstone, Zebra Stone

TEMPERAMENT (ANIMALS & CHILDREN)
Landscape Jasper, Green Apatite

TENSION
Aragonite, Fuchsite, Gypsum Desert Rose, Hypersthene, Bornite

THROAT
Angelite, Celestite, Copal, Blue & Rainbow Fluorite, Lazurite, Sodalite

THUNDER & LIGHTNING
Moss Agate, Hypersthene

THYROID/THYMUS
Angelite, Prehnite, Rutile Quartz, Dumortierite

TISSUES (BODY)
Yellow Jasper, Prehnite

TISSUE REPAIR
Angelite, Copper

TONSILS
Amber, Blue Lace Agate, Sodalite

TOXICITY
Yellow Apatite, Azurite, Celestite, Red Jasper

TRANSFORMATION
Charoite, Bloodstone, Purpurite,Selenite, Sapphire, Tiger's Eye

TRAVEL (SAFE)
Agate, Moonstone

TRAVEL SICKNESS
Malachite, Jet, Sapphire, Pink Tourmaline

TUMORS
Malachite, Bloodstone, Smoky Quartz

U

ULCERS
Sunstone, Lepidolite

URINARY TRACT/INFECTION
Amber, Blue Lace Agate, Red Calcite, Carnelian, Ruby

V

VERBAL EXPRESSION
Chrysacolla, Emerald, Blue Fluorite, Fulgarite, Moqui Marble

VARICOSE VEINS
Amber, Blue Lace Agate, Bloodstone

VERTEBRAE
Azurite, Axinite, Yellow Calcite,
Carnelian, Petrified Wood

VERTIGO
Lazurite

VISION
Unakite, Blue Tiger's Eye

VISION (NIGHT)
Amber, Fire Agate, Tiger's Eye

VITAMIN ABSORPTION
Apache Tears, Carnelian, Howlite,
Moonstone, Tiger Iron

VITAMIN B
Tiger Iron

VOCAL CHORDS
Sodalite, Stromatolite

VOMITING
Emerald, Lazurite

W

WALKING (INFANTS)
Yellow Agate

WARMTH (PHYSICAL)
Aragonite, Garnet

WARTS
Blue Lace Agate, Smoky Quartz

WATER (FEAR)
Blue Obsidian, Ocean Jasper

WATER PURIFICATION
Lodestone, Shungite

WEIGHT LOSS
Angelite, Apatite, Yellow Jasper

WHOOPING COUGH
Amber, Blue Lace Agate, Topaz

WISDOM
Apophyllite, Bismuth, Blue Agate,
Iolite, Sodalite

WORRY
Snakeskin Agate

WRISTS
Moss Agate, Dalmatian Jasper

WRITER'S BLOCK
Peruvian Agate, Carnelian, Diopside,
Vanadanite

X

X-RAYS (PROTECTION & RECOVERY)
Yellow Agate

Y

YIN YANG (BALANCE)
Rhodonite, Moqui Marbles

YOGA
Purple Agate, Herkimer Diamond,
Hypersthene, Lepidolite

SPECIMEN INDEX BY CLASSICAL ELEMENT

Classical elements refer to ancient beliefs inspired by natural observation of nature and of matter. In the western belief system there are four elements: earth *(solids)*, water *(liquid)*, air *(gas)* and fire *(plasma)*. The basis of these elements, as well as description and classification of chemical compounds and natural substances, can be traced back as far as the ancient Greek culture. Healing stones and crystals can be indicated to further connect ones self via the natural elements, to a desired outcome or effect.

The following list divides all healing crystals mentioned in this book, into their respective elements:

EARTH

Crazy Lace Agate
Moss Agate
Peruvian Agate
Tree Agate
White Agate
Yellow Agate
Amazonite
Chevron Amethyst
Anhydrite
Anyolite
Apache Tears
Apophyllite
Arsenic Orpiment
Arsenic Realagar
Axinite
Barite
Boji Stone
Bronzite
Blue Calcite
Calcite Fairy Stone
Orange Calcite
Red Calcite
Yellow Calcite
Celestite
Chalcedony
Chiastolite
Chrysacolla
Chrysoprase
Copal
Epidote

Feldspar
Galena
Fuchsite
Brecciated Jasper
Dalmatian Jasper
Fancy Jasper
Kambaba Jasper
Orange Jasper
Picture Jasper
Yellow Jasper
Zebra Jasper
Kyanite
Lazurite
Lepidolite
Lodestone
Magnesite
Malachite
Marcasite
Moqui Marble
Black Obsidian
Snowflake Obsidian
Silver Sheen Obsidian
Obsidian Glass
Onyx
Opal
Petrified Wood
Rhodocrosite
Rhodonite
Rutile Quartz
Smoky Quartz
Sardonyx
Septarian
Stromatolite

Tiger Iron
Black Tourmaline
Blue Lace Agate
Turritella Agate
Opal
Thulite

WATER

Blue Lace Agate
Turritella Agate
Anhydrite
Aquamarine
Azurite
Carnelian
Chrysoberyl
Dumortierite
Emerald
Blue Fluorite
Green Fluorite
Rainbow Fluorite
Herkimer Diamond
Iolite
Jade
Nephrite Jade
Larimar
Larvikite
Mookaite
Moonstone
Obsidian Glass
Opal
Pinolite
Prehnite

Prophecy Stone
Purpurite
Sapphire
Sodalite
Staurolite
Tanzanite
Wavellite

FIRE

Fire Agate
Bloodstone
Bumblebee
Brown Calcite
Cinnabar
Citrine
Copper
Creedite
Garnet
Gold
Hematite
Desert Jasper
Leopardskin Jasper
Ocean Jasper
Red Jasper
K2 Stone
Obsidian Glass
Opal
Peacock Ore
Pyrite
Clear Quartz
Tangerine Quartz
Tourmilinated Quartz
Rainforest Rhyolite
Ruby
Serpentine
Sulfur
Sunstone
Tektite
Blue Tiger's Eye
Gold Tiger's Eye
Red Tiger's Eye
Pink Tourmaline
Turquoise

Unakite
Vanadanite
Vesuvianite

AIR

Blue Lace Agate
Crazy Lace Agate
Moss Agate
Peruvian Agate
Tree Agate
White Agate
Yellow Agate
Amazonite
Chevron Amethyst
Apache Tears
Apophyllite
Axinite
Barite
Boji Stone
Bronzite
Blue Calcite
Orange Calcite
Red Calcite
Yellow Calcite
Carborundum
Celestite
Chalcedony
Chiastolite
Chrysacolla
Chrysoprase
Copal
Epidote
Feldspar
Galena
Fuchsite
Brecciated Jasper
Dalmatian Jasper
Fancy Jasper
Kambaba Jasper
Orange Jasper
Picture Jasper
Yellow Jasper
Zebra Jasper

Kyanite
Lazurite
Lepidolite
Lodestone
Malachite
Moqui Marble
Black Obsidian
Snowflake Obsidian
Silver Sheen Obsidian
Obsidian Glass
Onyx
Opal
Petrified Wood
Rhodocrosite
Rhodonite
Rutile Quartz
Smoky Quartz
Sardonyx
Septarian
Stromatolite
Tiger Iron
Black Tourmaline
Sugilite
Zircon

CHAKRA INDEX TO STONES

Many people find that certain crystals and stones are useful for healing, charging or realigning the bodily chakras. In some cases the color of a stone may correspond to the traditional color interpretations of the chakras. In other cases, stones are chosen for their particular effect on one or all chakras. For realigning the entire body, it has been suggested that you choose and use one stone for each chakra. As with all crystal healing, use of a good clear quartz crystal, prism, or wand, can amplify the energy of the other stones. Following is an index of crystals that may be useful for each chakra. Please remember that not every stone will correspond in color to the chakra for which it is most suited. The lists that follow, divide all specimens from this book into their corresponding chakras.

THE TRADITIONAL COLORS THAT CORRESPOND TO THE CHAKRAS:

BASE CHAKRA	Red
SACRAL CHAKRA	Orange
SOLAR PLEXUS CHAKRA	Yellow
HEART CHAKRA	Green
THROAT CHAKRA	Blue
THIRD EYE CHAKRA	Indigo
CROWN CHAKRA	Violet or Gold *(Can substitute clear Quartz)*

STONES FOR THE CHAKRAS

BASE

Calcite Fairy Stone
Cinnabar
Schalenblende
Snakeskin Agate
Turritella Agate
Apache Tears
Bloodstone
Bronzite
Carnelian
Copper
Galena
Hematite
Brecciated Jasper
Dalmatian Jasper
Leopardskin Jasper
Red Jasper
Zebra Jasper
Jet
Kyanite
Labradorite
Black Obsidian
Snowflake Obsidian
Onyx
Smoky Quartz
Ruby
Sardonyx
Shiva Lingam
Red Tiger's Eye
Black Tourmaline

SACRAL

Blue Lace Agate
Botswana Agate
Bumblebee
Fire Agate
Chalcedony
Chiastolite
Citrine
Epidote
Green Fluorite

Gold
Fancy Jasper
Orange Jasper
Prehnite
Star Mica
Tangerine Quartz
Sunstone
Tiger's Eye
Topaz

SOLAR PLEXUS

Pinolite
Arsenic Orpiment
Arsenic Realgar
Snakeskin Agate
Turritella Agate
Apache Tears
Bloodstone
Bronzite
Carnelian
Copper
Galena
Hematite
Brecciated Jasper
Dalmatian Jasper
Leopardskin Jasper
Red Jasper
Zebra Jasper
Jet
Kyanite
Labradorite
Black Obsidian
Snowflake Obsidian
Onyx
Smoky Quartz
Ruby
Sardonyx
Shiva Lingam
Red Tiger's Eye

HEART CHAKRA

Staurolite
Vesuvianite
Wavellite
Aventurine
Red Calcite
Charoite
Chrysoberyl
Chrysoprase
Diamond
Diopside
Emerald
Blue Fluorite
Garnet
Hematite
Jade
Nephrite Jade
Kambaba Jasper
Larvikite
Picasso Jasper
Malachite
Mookaite
Green Obsidian
Peridot
Petalite
Petrified Wood
Green Quartz
Rose Quartz
Rainforest Rhyolite
Rhodocrosite
Rhodonite
Ruby Fuchsite
Serpentine
Turquoise
Umakite
Yellow Jasper

STONES FOR THE CHAKRAS CON'T

THROAT CHAKRA

Larimar
Iolite
Thulite
Blue Agate
Amazonite
Anhydrite
Blue Apatite
Aquamarine
Barite
Blue Calcite
Brown Calcite
Celestite
Chrysacolla
Dumortierite
Noreena Jasper
Ocean Jasper
Lazurite
Blue Obsidian
Sapphire
Shattuckite
Sodalite
Tanzanite
Blue Tiger's Eye

THIRD EYE

Amethyst
Anyolite
Chevron Amethyst
Violet Amethyst
Apophyllite
Azurite
Creedite
Purple Fluorite
Fulgarite
Hypersthene
Lepidolite
Lodestone
Marcasite

Moonstone
Nuummite
Silver Sheen Obsidian
Pietersite
Phantom Quartz
Scolecite
Septarian
Stilbite
Tektite
Thulite
Vanadanite
Yellow Agate

CROWN

Moss Agate
Peruvian Agate
Pink Botswana Agate
Purple Agate
Snakeskin Agate
Tree Agate
White Agate
Blue Aragonite
Orange Aragonite
White Aragonite
Iceland Spar
Rainbow Fluorite
Herkimer Diamond
K2 Stone
Magnesite
Snowflake Obsidian
Opal
Prophecy Stone
Purpurite
Pyrite
Clear Quartz
Tourmilinated Quartz
Selenite
Sugilite
Stromatalite
Pink Tourmaline

BIRTHSTONES BY MONTH

The following is a list of traditional gem birthstones followed by their alternative semi-precious stones, corresponding to each month of the year.

JANUARY	Garnet, Rose Quartz, Peacock Ore
FEBRUARY	Amethyst, Onyx, Black Tourmaline
MARCH	Aquamarine, Heliotrope, Jade
APRIL	Diamond, Clear Quartz
MAY	Emerald, Chrysoprase, Aventurine
JUNE	Pearl, Moonstone, Howlite
JULY	Ruby, Carnelian, Tiger's Eye
AUGUST	Peridot, Sardonyx, Amazonite
SEPTEMBER	Sapphire, Lapis Lazuli, Sodalite
OCTOBER	Opal, Tourmaline, Fluorite
NOVEMBER	Topaz, Citrine, Obsidian
DECEMBER	Zircon, Tanzanite, Jasper

ASTROLOGICAL INDEX TO STONES BY ZODIAC

Certain healing stones and crystals correspond to signs of the zodiac. This is not to be confused with birthstones, which traditionally are linked to calendar months for the purposes of gifting and jewelry, not specifically for healing purposes.

ARIES STONES

Anyolite
Apache Tears
Aquamarine
Aventurine
Axinite
Bloodstone
Magnesite
Orange Calcite
Yellow Calcite
Citrine
Diamond
Emerald
Garnet
Hematite
Jade
Brecciated Jasper
Kyanite
Lodestone
Ruby
Tektite

TAURUS STONES

Calcite Fairy Stone
Carnelian
Chrysacolla
Copper
Iolite
Lazurite
Preseli Bluestone
Rhodonite
Sapphire
Septarian
Black Tourmaline
Pink Tourmaline
Zircon

GEMINI STONES

Arsenic Realgar
Blue Agate
Crazy Lace Agate
Fire Agate
K2 Stone
Moss Agate
Peruvian Agate
Pink Botswana Agate
Purple Agate
Thulite
Tree Agate
Turritella Agate
Blue Apatite
Yellow Apatite
Celestite
Diopside
Epidote
Howlite
Leopardskin Jasper
Onyx
Rutile Quartz
Sardonyx
Selenite
Serpentine

CANCER STONES

Blue Calcite
Brown Calcite
Chalcedony
Chiastolite
Chrysoprase
Cinnabar
Pinolite
Red Jasper
Silver
Star Mica
Moonstone

LEO STONES

Arsenic Orpiment
Bronzite
Bubble Bee
Chrysoberyl
Copal
Dumortierite
Rainbow Fluorite
Gold
Iceland Spar
Desert Jasper
Fancy Jasper
Picasso Jasper
Picture Jasper
Yellow Jasper
Zebra Jasper
Labradorite
Mookaite
Peridot
Petalite
Rhodochrosite
Sunstone
Topaz

VIRGO STONES

Amazonite
Amethyst
Violet Amethyst
Astrophyllite
Larvikite
Purpurite
Red Calcite
Purple Fluorite
Fulgarite
Vanadanite
Yellow Agate

LIBRA STONES

Apophyllite
Herkimer Diamond
Hypersthene
Jade Nephrite
Noreena Jasper
Lepidolite
Malachite
Moqui Marble
Prehnite
Staurolite
Sugilite
Sulfur

SCORPIO STONES

Botswana Agate
Snakeskin Agate
Boji Stone
Charoite
Green Fluorite
Galena
Dalmatian Jasper
Kambaba Jasper
Black Obsidian
Green Obsidian
Marcasite
Shiva Lingam
Turquoise
Unakite
Yellow Agate

SAGITTARIUS STONES

Azurite
Barite
Herkimer Diamond
Nuummite
Yellow Obsidian
Peacock Ore
Prophecy Stone
Rainforest Rhyolite
Sodalite
Vesuvianite

CAPRICORN STONES

Blue Aragonite
Orange Aragonite
White Aragonite
Blue Fluorite
Ocean Jasper
Jet
Schalenblende
Scolecite
Snowflake Obsidian
Blue Tiger's Eye
Gold Tiger's Eye
Red Tiger's Eye

AQUARIUS STONES

Anhydrite
Carborundum
Feldspar
Fuchsite
Ruby Fuchsite
Prehnite
Wavellite

PISCES STONES

Blue Lace Agate
Larimar
Blue Obsidian
Opal
Stromatolite
Silver Sheen Obsidian
Tanzanite

HOUSEHOLD PROTECTION

There are many methods of using crystals and stones for healing and protecting a household. Some involve placing a grid of crystals in strategic and energetically related locations throughout your home *(see next page)*. Other methods are more specific in location, and have practical functions for each room in the house, and have been suggested to have beneficial results. Following is a list of uses for placing crystals in your home. Keep in mind that any stones, rocks, or crystals that you may now have displayed in your house, are already part of your personal healing arsenal by virtue of their beauty and energy. Use and enjoy them all!

KITCHEN	Quartz for healthful cooking
DINING ROOM	Citrine for good digestion
BEDROOM	Chrysocolla for vital energy
CHILD'S BEDROOM	Kyanite or howlite for happy dreams
LIVING ROOM	Fluorite for harmony
FAMILY ROOM	Geode for communication
HOME OFFICE	Amethyst for protection from EM radiation
NEAR PHONES	Rhodonite for patience with others
HALLWAYS	Labradorite for transitions
ENTRY OR FOYER	Carnelian for protection
GARAGE	Agate for safe travel
WORKSHOP	Bloodstone for safety
BATHROOM	Quartz for purification
FRONT DOOR	Hematite to repel negative energy
BACK DOOR	Obsidian to protect from burglary
GARDEN	Mookaite for healthy growth

HOME AND HOUSEHOLD GRID

The following is a general guideline for creating a crystal grid in the home for protection, spiritual cleansing, positive energy, and other beneficial effects. As with any recommendations for crystal use, substituting these suggested stones to any others with which you feel a special affinity, is completely acceptable and encouraged. You should make your family members aware of the grid you have created, simply so they do not inadvertently disturb the placement of your crystals and so they understand the purpose of the grid. Checking and cleansing the grid-stones need not be done more than once a year. Set your grid-stones as follows:

NORTH	Clear Quartz for success and good decision making as a family
NORTHEAST	Carnelian for tolerance of visitors and for always keeping an open mind
EAST	Jade for long, healthful life and wisdom
SOUTHEAST	Citrine for wealth and prosperity, and as a warning against greed and excess
SOUTH	Amethyst or quartz for a positive family reputation and improving good fortune
SOUTHWEST	Rose quartz for harmonious relationships especially among siblings
WEST	Moonstone for fertility, health, and protection of children
NORTHWEST	Turquoise for mentoring and support of each other
CENTRAL	Family favorites should be placed thoughtfully throughout the home

CARE AND CLEANSING OF CRYSTALS

Since many uses for crystals involve protection or healing by removing or absorbing negative energy and influence, it is recommended to cleanse and purify your stones and crystals occasionally. There is no agreed upon method for doing so, but many people choose to wash them in clear water and sometimes mild soap, and/or recharge them with sunlight and fresh air. Some of the more colorful, translucent or delicate stones can fade in direct sunlight, so moonlight recharging or water washing may be preferable.

Some stones that should not be displayed or energized in direct sunlight are amethyst, apatite, celestite, citrine, opal, peacock ore, smoky quartz, and any of the beryl gemstones *(emerald, heliodor, aquamarine, morganite)*.

Stones that should never be cleansed or immersed in water are azurite, apatite, apophyllite, calcite, carnelian, gypsum desert rose, halite, hematite, labradorite, lepidolite, lodestone *(magnetite)*, malachite, mica, opal, pyrite, and turquoise.

Stones that many people feel do not need cleansing include bloodstone, carnelian, clear quartz, jet, selenite and kyanite.

In general, if you are handling your crystals during illness or have let others handle them, it is a good idea to wash them occasionally. It is a kind and worthy action to occasionally offer some of your stones and crystals to others, when trying to help them with their own healing or meditational needs. However, any stone with which you feel a special affinity or personal attachment need not be handled by others. It is quite alright to keep your special stones for just yourself!